PHYTOREMEDIATION: GREEN SOLUTION TO POLLUTION

PHYTOREMEDIATION: GREEN SOLUTION TO POLLUTION

Edited by

Dr. Bhupendra Pushkar

Delve
PUBLISHING

www.delvepublishing.com

Phytoremediation: Green Solution to Pollution

Dr. Bhupendra Pushkar

Delve Publishing
224 Shoreacres Road
Burlington, ON L7L 2H2
Canada
www.delvepublishing.com
Email: orders@arclereducation.com

ISBN: 978-1-77469-867-9 (Hardcover)

ABOUT THE EDITOR

Dr. Bhupendra Pushkar was born in Mumbai, India. Since 2008 he is working as Assistant professor at University of Mumbai, India. He pursued his B.Sc. in Microbiology in 2000. He has obtained his M.Sc. degree in Biotechnology in 2002 from University of Mumbai. He pursued his further studies and completed his Doctorate degree in 2010. His area of research broadly includes environmental pollution remediation using biological science. He is currently working on heavy metal bioremediation, PAH bioremediation, plastic degradation, plant tissue culture etc. He is actively involved in the bioremediation research work and trying to develop system for river water cleaning. He has published various papers in national as well as international journals and currently guiding three research students. He is Co-Editor in Chief of a journal and reviewer in many journals.

TABLE OF CONTENTS

LIST OF FIGURES

PREFACE

The environment is made up of air, water, and soil, which are essential for all biological processes and sustain the growth of all living organisms on Earth. The global spread of toxic organic and/or inorganic pollutants into water, soil, and air due to anthropogenic sources and natural disasters causes a number of issues for the environment and human health. Depending on the type of pollution, pollutants have a variety of adverse consequences on the environment and human health. In order to mitigate the negative impacts on human health and protect the environment for future generations, there is a vital need for the remediation of environmental toxins from soil, water, and air. For decades, a variety of techniques based on physical, chemical, as well as biological means have been available for the remediation of contaminated sites. However, these techniques are energy and money-intensive, cause secondary air, soil, and groundwater pollution, and have the potential to kill the microbial communities that break down pollutants. Because of this, the preservation of our natural environment necessitates the development of sustainable environmental remediation technologies that promise to restore contaminated sites in an efficient and ecologically acceptable way for human rehabilitation and agricultural production.

The application of phytotechnologies is still progressing steadily. More ecologists, engineers, and government officials are now aware of phytoremediation as a green, affordable, and workable solution to some of the many environmental problems. Emerging pollutants like heavy metals, volatile organic carbons, pesticides, personal care products, medicines, organics, NPs, and pathogenic bacteria have been known to disturb water and the environment over the decades. These contaminants come from a variety of human endeavors, so readers ought to be naturally interested in and drawn to them. A thorough review of phytoremediation, including its processes and applied features, is offered in order to comprehend the plant-based approaches for reducing water and soil pollution from these emerging environmental pollutants, particularly heavy metals. Furthermore, there are limitations imposed by a critical assessment of the benefits and drawbacks of phytoremediation technology that must be overcome. This book describes the developments in phytoremediation around the world.

This book serves as a guide for researchers and graduate students interested in phytoremediation and its applications.

Introduction

The issue of energy and the environment, specifically the trends toward global warming brought on by the emission of greenhouse gasses and the energy crises due to the widening gap between the limited supply and demand for petroleum and other fossil fuels, is one of the most pressing issues faced by humanity now. Soil evacuation and *ex-situ* treatment by burning or secured land-filling are the main solutions for remediating soils contaminated by explosive compounds. However, this approach is incredibly expensive, harmful to the environment, and completely impractical. Cost-effective phytoremediation and eco-friendly techniques were tested on several bacterial strains and plant types. It has been demonstrated that a variety of fungi, yeast, bacteria, and other microorganisms found in the root zone of higher plants degrade organics like explosives, fuels, and solvents. Willow and Poplar are some examples of plants that have been widely used to clean up the soil. Biofuels are a type of bioenergy that may be converted into fuel. In the twenty-first century, bio-fuels made from various types of biomasses are seen as viable alternatives to natural oil since they are both environmentally and economically sound.

Chemical compounds are produced and used imprecisely as a result of rapid industrialization, rising urbanization, and altered agricultural methods. As a result, chemicals that are harmful to both the environment and human health have significantly contaminated the ecosystem. Heavy metals and organic contaminants are currently co-contaminating a lot of sites. As a result, the removal of contaminants has emerged as a significant issue that calls for a range of strategies to find appropriate solutions. For *in situ* decontamination of soil and water, phytoremediation is important. The use of plants to remove contaminants or render them harmless through physical, chemical, and biological processes is a low-cost and ecologically acceptable approach. One of the main issues that substantially reduce agricultural output across the globe is soil salinization. More than 500 million hectares of land are not being exploited effectively for food production because of salt. To address the current issues of global food security, it is necessary to find ways to make saline soils better so that they can support highly productive and significant land-use systems. For example, in order to meet the plants' needs for nitrogen, cyanide must be transported and assimilated by the plants through very intricate physiological mechanisms. Various plant species, such as trees, grasses, and aquatic plants, have different mechanisms for transporting and metabolizing various chemical forms of cyanide.

CHAPTER 1

Introduction to Phytoremediation

CONTENTS

1.1 INTRODUCTION

Pesticides, medications, petroleum hydrocarbons, PAHs, PCBs, and other persistent organic pollutants (POPs) are known global environmental problems. Cost-effective solutions have been created and assessed for the decontamination of soil and water resources in order to address this issue. A potential approach called phytoremediation (Figure 1.1) employs plants and the microbes in their rhizosphere to accumulate, alter, or eliminate organic and inorganic pollutants from soils, sediments, surface or groundwater, wastewater, and even the atmosphere. In fact, due to their sedentary nature, plants have developed a variety of defence mechanisms against hazardous substances in their surroundings (Salt, Smith, and Raskin, 1998). This makes their application in cleaning up contaminated land and water more practical than physical and chemical methods because they have a variety of pollutant attenuation mechanisms. A variety of decontamination techniques are used in phytoremediation, which is currently used to remediate several kinds of organic xenobiotics, such as petroleum hydrocarbons, chlorinated solvents, polycyclic aromatic hydrocarbons, pesticides, explosives, medicinal chemicals, and their metabolites. Phytotechnologies come in a variety of forms, such as the treatment of artificial wetlands.

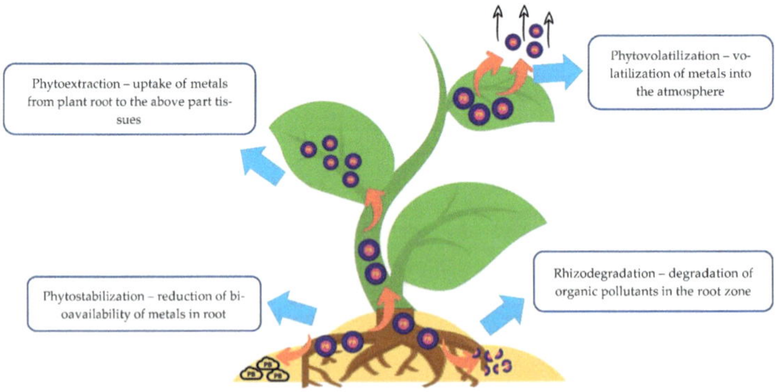

Figure 1.1. Phytoremediation.

Source: *Cleophas FN, Zahari NZ, Murugayah P, Rahim SA, Mohd Yatim AN. Phytoremediation: A Novel Approach of Bast Fiber Plants (Hemp, Kenaf, Jute and Flax) for Heavy Metals Decontamination in Soil—Review. Toxics. 2023; 11(1):5. https://doi.org/10.3390/toxics11010005*

Constructed wetlands for wastewater treatment, (Figure 1.2) which have been successfully employed to remove a variety of organic xenobiotics, merit special attention within the field of phytotechnologies. The water supply and drinking water quality are under stress due to demographic pressures and rising living standards. Some of this pressure may be reduced with the use of demand control and water and wastewater reuse. New compounds of concern, such as hazardous substances and endocrine disruptors, which frequently flow through wastewater treatment plants (WWTPs) without being effectively removed, provide new issues in wastewater management. Such compounds have the potential to have negative effects on human health and aquatic ecosystems (Mirzaee, ZakeriNia, and Farasati, 2021).

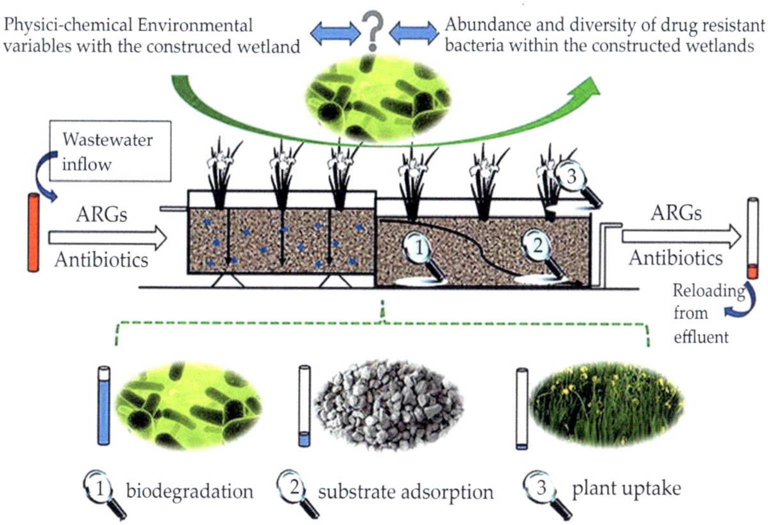

Figure 1.2. Constructed wetlands for wastewater treatment.

Source: *Oscar Omondi, D., and Caren Navalia, A. (2021). Constructed Wetlands in Wastewater Treatment and Challenges of Emerging Resistant Genes Filtration and Reloading. IntechOpen. doi: 10.5772/intechopen.93293*

1.2 POLLUTANTS

In the modern world, tens of thousands of different organic substances are used. Pesticides, organic solvents, explosives, dyes, phenols, petroleum hydrocarbons, and a new class of emerging contaminants called medicines are among these substances. Numerous of these organic contaminants are environmentally persistent and may have unfavorable ecotoxicological effects. Pharmaceutical contamination of the aquatic environment has just lately come to light, even though some of these kinds of pollutants have long been recognized to exist in the environment and have hazardous effects that have been thoroughly investigated (Kurade et al., 2021). This is because analytical methodologies and instrumentation with detection and quantification limits to permit the environmental monitoring of these substances have only recently become available. As a result, these pollutants are typically found in that medium at very low concentration levels. According to several studies, the ecotoxicity associated with this class of pollutants is also largely unassessed. However, there is potential for chronic effects caused by long-term exposure as well as for cumulative and even synergistic action. Nevertheless, pharmacologically active chemicals that are necessary for the diagnosis, treatment, or prevention of diseases are used in increasing amounts each year in human and veterinary medicine. Pharmaceuticals and their metabolites are regularly introduced into the sewage system through the discharge or disposal of unwanted or expired items. Many of these substances are eventually released into the environment as a result of ineffective treatment in WWTPs. This is thought to be the primary method by which medicines contaminate the aquatic environment. Remains of lipid-regulating drugs, analgesics and anti-inflammatory drugs, antibiotics, hormones, antidiabetics, neuroactive compounds, and beta-blocker drugs have all been found in wastewaters, surface waters, ground waters, and even drinking waters worldwide over the past few years in numerous monitoring studies.

Despite the occasionally high removal efficiencies realized, these methods are typically not cost-effective on a large scale. It remains a major objective to find appropriate technology that is more efficient and has affordable operating and maintenance costs. Constructed wetlands systems (CWS), a phytoremediation technique for wastewater treatment that has grown in popularity over the past few decades, take advantage of how wetlands can purify water while attempting to improve the processes involved (Mant et al, 2006). These systems are increasingly being used as an alternative to traditional wastewater treatment methods or as a secondary or tertiary treatment step in WWTPs. Their most appealing traits include low cost and low maintenance. For the removal of bulk pollutants such as suspended particles, organic matter, pathogens, and nutrients, CWS may be viewed as a more developed method nowadays. The removal of more specific substances, for which traditional wastewater treatment technologies are ineffective, is currently the area of focus. In fact, CWSs are now being employed more frequently for the degradation of certain pollutant types, including organic xenobiotics, and new challenges, like the removal of pharmaceuticals and other micropollutants, have been arising, presenting new environmental issues to be tackled. Numerous novel chemical compounds have been released into the environment as a result of the explosive growth of chemical and agrochemical industries during the past century. A wide variety of organic compounds are now used by people daily, and many of these are frequently found in numerous environmental monitoring studies. A group of dangerous organic substances, notably xenobiotics. These are synthetic molecules made by humans that do not naturally occur in biological systems, which are referred to as persistent organic pollutants (POPs), (Figure 1.3) and have come under increased scrutiny over the past few decades (Lenoir et al, 2016). POPs are chemicals that remain in the environment, bioaccumulate up the food chain, and constitute a concern to the environment and human health since they can be hazardous to biotic communities when present at levels above the baseline. These POPs include petroleum hydrocarbons, polycyclic aromatic hydrocarbons, chlorinated solvents, explosives, dyes, phenols, and pesticides.

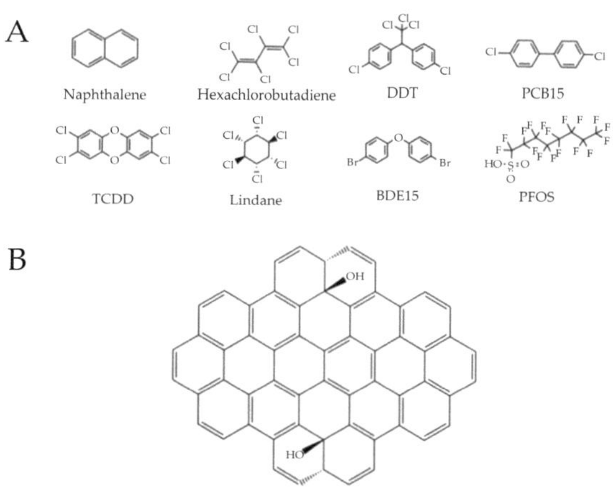

Figure 1.3. Persistent organic pollutants.

Source: *Wu Q, Zhang R, Wang X, Li Y. A Theoretical Study of the Interactions between Persistent Organic Pollutants and Graphene Oxide. International Journal of Environmental Research and Public Health. 2022; 19(18):11340. https://doi.org/10.3390/ijerph191811340*

1.3 XENOBIOTICS

Organic xenobiotics (Figure 1.4) can be dispersed into many areas of the environment, transported over short or long distances, and subjected to a range of reactions and transformations (Schröder et al, 2008). Due to these numerous conflicting interactions, it is impossible to forecast how such pollutants would behave, and in many circumstances, it is also difficult to evaluate how they will affect the environment. However, the international community has repeatedly called for urgent global action to be taken to reduce the release of these chemicals due to the established possibility of long-range transport of these substances to regions where they have never been used or produced and the resulting threats they may pose to the environment of the entire globe. Point sources of wastewater discharge, incorrect usage and disposal, or unintentional environmental releases of organic compounds are all causes of soil and surface or groundwater pollution. Numerous techniques and technologies have been created recently for the treatment of wastewater or the restoration of contaminated areas. Advanced oxidative processes, activated carbon adsorption, membrane filtration, and membrane bioreactors are a few of the advanced wastewater treatment technologies that have been studied for the removal of POPs. Overall, a number of methods have been created and put to use for decontaminating soils and natural streams. Detergent washing, steam or hot air stripping, flooding and other in situ techniques are a few examples. Ex situ methods involve excavating the contaminated soil or pumping the liquid, treating it with chemical oxidation, solvent extraction, adsorption, thermal desorption, etc., and then reintroducing the cleaned soil or liquid to the original location (Pittarello et al, 2017).

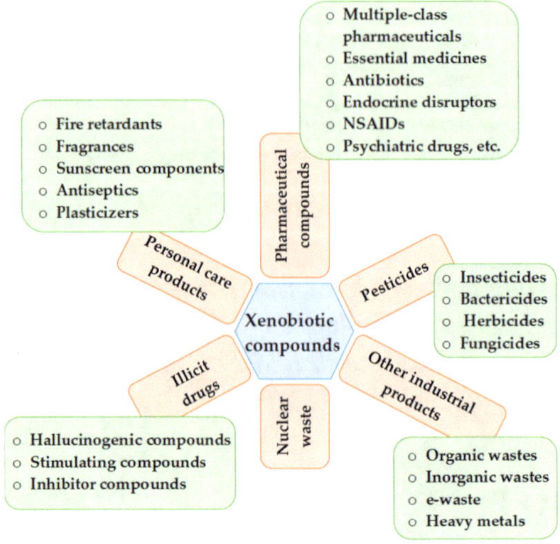

Figure 1.4. Organic xenobiotics.

Source: *Štefanac T, Grgas D, Landeka Dragičević T. Xenobiotics—Division* *and Methods of Detection: A Review. Journal of Xenobiotics. 2021; 11(4):130-141. https://doi.org/10.3390/jox11040009*

Although these methods can sometimes achieve high removal efficiencies, they are not generally employed for cost-effectiveness reasons. As a result, there is a rising need for alternative POP removal techniques for soils, natural waterways, and wastewaters that are more effective and have affordable operation/maintenance costs. Many POPs have been removed from contaminated soils, waterways, and wastewater using phytotechnologies. This method aims to take advantage of plants' capacity to absorb, concentrate, or metabolize organic xenobiotics as well as their ability to exude substances from their roots that facilitate compound biotransformation and microbial destruction.

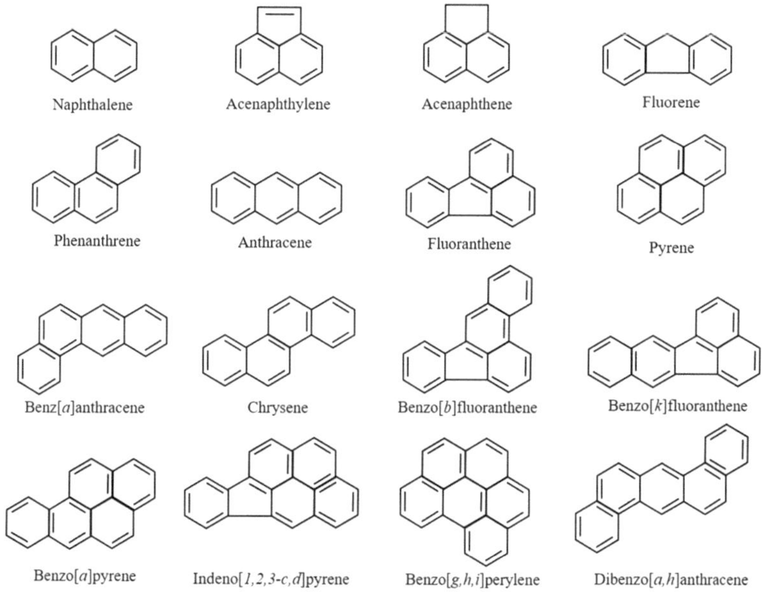

Figure 1.5. Chemical structure of the 16 representative polycyclic aromatic hydrocarbons (PAHs) as decided upon by the United States Environmental Protection Agency (US EPA)

Source: *Honda M, Suzuki N. Toxicities of Polycyclic Aromatic Hydrocarbons for Aquatic Animals. International Journal of Environmental Research and Public Health. 2020; 17(4):1363. https://doi.org/10.3390/ijerph17041363*

The use of various phytotechnologies, such as CWS, has gained popularity, especially for the removal of organic xenobiotics from raw wastewater and WWTP effluents. The broad class of several hundred chemical compounds known as polycyclic aromatic hydrocarbons (PAHs) (Figure 1.5) is distinguished by the presence of two or more fused aromatic rings (Sarma, Nava, and Prasad, 2019). These substances are significant environmental pollutants due to their pervasiveness in the environment and the fact that some of them are regarded as hazardous materials due to their toxicity and propensity for mutation or cancer. Due to this, a number of them are listed on the priority pollutant list by the Environmental Protection Agency (EPA) of the United States and the European Union. Anthropogenic and natural sources both release PAHs into the environment (McIntyre, 2003). These chemicals can also be produced by bacteria, fungi, plants, and animals. They are created when nearly any organic material burns incompletely such as fossil fuels, in forest fires, during volcanic eruptions, in car exhaust, etc. Significant PAH emissions from crude oil and other petroleum-based products are also harmful to the environment.

1.4 POLYCYCLIC AROMATIC HYDROCARBONS (PAHS)

The dangers posed by PAHs can be avoided by using standard techniques that involve the removal, modification, or isolation of the pollutant. These methods include digging up contaminated soil and burning it or containing it, as well as thermal desorption, solvent extraction, and chemical oxidation. The contaminant is frequently transferred from one phase to another by these costly technologies. Phytotechnologies have been utilized to remove PAHs from contaminated soils, waterways, and wastewater. Because some PAHs are not readily absorbed by plants due to their physicochemical properties such as water solubility, they appear less suitable for phytodegradation. The most complex naturally occurring mixture of organic molecules, crude oil is a lipophilic mixture made up of more than 17000 different chemical components (Pradhan et al, 1998). The primary components are PAHs, cycloalkanes, and alkanes. Total petroleum hydrocarbon (TPH) pollution of the environment results from both anthropogenic and natural sources and is exacerbated by the widespread use of so many petroleum-based products in contemporary civilization. Offshore oil production, maritime transportation, atmospheric or aerial depositions from coal and gas flare combustion, direct ocean dumping, coastal, municipal, and industrial wastes, and runoff are examples of human-mediated sources of TPHs. Only a relatively tiny portion of TPHs have adequate toxicity data. The health impacts of various fractions, such as the light aromatics and, in particular, the BTEX (benzene, toluene, ethylbenzene, and xylenes) fraction, (Figure 1.6) can be well defined based on their constituents or representative molecules.

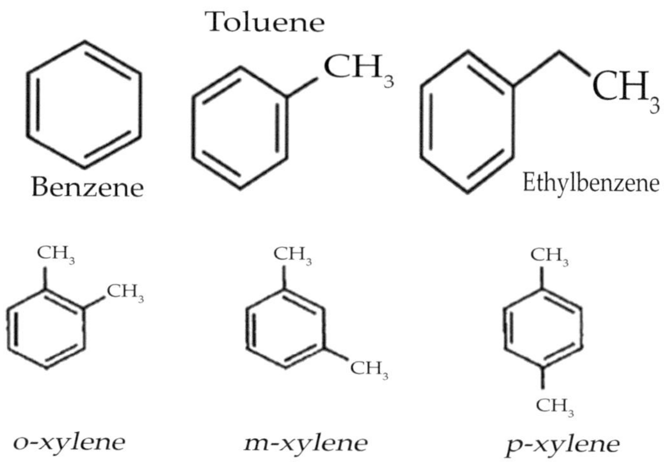

Figure 1.6. BTEX compound structures.

Source: *Fayemiwo, OM, Daramola, MO, and Moothi, K. (2017). BTEX compounds in water - future trends and directions for water treatment. Water SA, 43(4), 602-613. https://dx.doi.org/10.4314/wsa.v43i4.08*

Some petroleum-based fuels often contain these monoaromatic hydrocarbons, which are characteristic components of TPHs wastewaters. Because it is the most hazardous and recognised human carcinogen, benzene is the BTEX chemical that should cause the most concern (Collins, Laturnus, and Nepovim, 2002). Since the benzene ring has a high degree of thermodynamic stability and is a highly common chemical structure in nature, many aromatic compounds are significant environmental contaminants because of their long persistence in the environment. Thus, benzene pollution is a serious issue. This hydrocarbon is poorly biodegradable in the absence of oxygen and is highly soluble and mobile, especially in ground and surface waters.

1.5 TECHNIQUES FOR TPH DECONTAMINATION

Techniques such as soil solidification which binds hydrocarbon to the soil, flooding, excavating contaminated soil and subjecting it to chemical oxidation or incineration, washing with detergent, solvent extraction, adsorption, etc. have all been tried as decontamination methods for soils and waters that have been contaminated with TPHs. Nevertheless, these approaches to cleaning up polluted regions are typically costly. The application of phytotechnologies for soil and natural water remediation has grown in recent years as a low-cost alternative to traditional methods. Particularly, the use of CWS for the treatment of wastewater that has been contaminated with TPHs has been viewed as an alternative solution that has frequently been proven to be effective. The term "chlorinated solvents" refers to a class of organic compounds with one or more chlorine atoms, which includes polychlorinated biphenyls as well as derivatives of methane, ethane, and ethene (PCBs). Chlorinated solvents are frequently used in the production of polymers, dry cleaning, degreasing processes, and as chemical intermediates. The three most common chlorinated solvents that are pollutants in the environment are trichloroethylene (TCE), perchloroethylene (PCE), and polychlorinated biphenyls (PCBs) Mackova et al., 2006) (Figure 1.7).

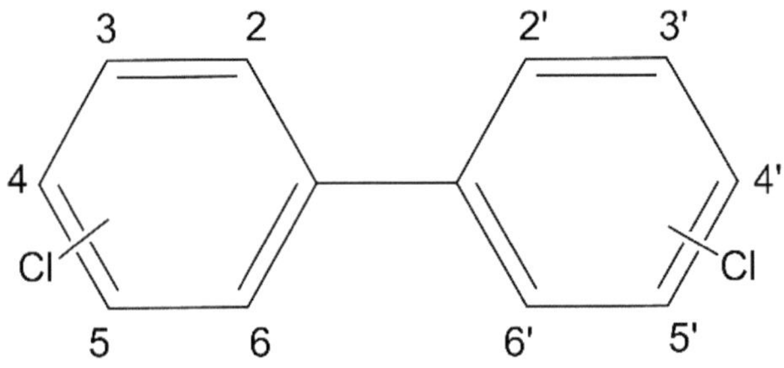

$C_{12}H_{10-x}Cl_x$, where x= 1–10

Figure 1.7. General structure and formula of PCBs. Chlorine atoms can replace hydrogens in different positions of the aromatic rings.

Source: *Montano L, Pironti C, Pinto G, Ricciardi M, Buono A, Brogna C, Venier M, Piscopo M, Amoresano A, Motta O. Polychlorinated Biphenyls (PCBs) in the Environment: Occupational and Exposure Events, Effects on Human Health and Fertility. Toxics. 2022; 10(7):365. https://doi.org/10.3390/toxics10070365*

TCE is a suspected carcinogen that predominantly affects the kidneys, liver, lungs, and heart rate. It is used as a metal cleaning agent and in speciality adhesives. Contrarily, PCE, which is also used to clean metal and in dry cleaning, is not considered a carcinogen but has been linked to central nervous system issues as well as skin, eye, and upper respiratory system irritation. PCBs are synthetic oils that are primarily utilized as coolants and/or insulators but do not readily react at room temperature. The International

Agency for Research on Cancer and the EPA have classed them as probable carcinogens. Because some by-products of PCB burning at low temperatures are extremely hazardous and carcinogenic, PCB contamination is a concern for the environment.

Except vinyl chloride, the majority of chlorinated solvents have densities that are higher than those of water and are only marginally soluble in water. When these two factors come together, dense non-aqueous phase liquid deposits are created. These deposits serve as a steady, slow-releasing source of chlorinated solvents. They typically take a long time to remove and tend to stay in the environment for extended periods of time as a result. Natural attenuation, soil vapour extraction, air sparging, and pump and treat techniques are examples of traditional ways of cleaning up chlorinated solvent pollution. Rhizodegradation, phytodegradation, phytovolatilization, and hydraulic control using hybrid poplar and willow trees are phytoremediation mechanisms that have been successful in containing and/or remediating chlorinated solvents, the use of CWS for the treatment of this class of POPs has also shown to be an effective low-cost approach. Explosives fall into one of three categories: nitroaromatics, nitramines, or nitrate esters. An international environmental issue is the pollution of the environment caused by explosives, particularly nitroesters and nitroaromatics. The majority of contaminated sites are found at ammunition factories and other locations where these compounds were handled because the manufacturing, storage, testing, and improper waste disposal of explosive chemicals are the main causes of soil and water contamination with explosives. This involved burning explosives at ordnance disposal sites, artillery ranges, evaluation facilities, and army stores (Hannink, Rosser, and Bruce, 2002).

1.6 EXPLOSIVES

A common nitroaromatic component of explosives is TNT. TNT may be stored for a long time and is stable in its refined state. It is not particularly sensitive to impacts. Alkalis easily react with it to produce unstable chemicals that are extremely susceptible to heat and impact. TNT exposure has health impacts such as anemia, poor liver function, irritated skin, and cataracts. A common nitramine used as an explosive and a component of plastic explosives is RDX. When RDX is consumed or inhaled in excessive doses, it can result in seizures (Panz and Miksch, 2012). There are no known long-term health consequences of low-level RDX exposure on the neurological system. The nitramine HMX bursts violently when heated up. Rocket fuel, plastic explosives, and nuclear devices all use it. There haven't been enough studies done on how HMX affects both human and animal health. Traditional techniques used to remove explosives pollution from soil and groundwater include incineration, landfilling, and pump-and-treat systems. These methods are expensive, and the production of ash can pollute the air. In constructed wetlands, different types of plants have been used in phytoextraction, phytodegradation, and phytostabilization processes that have been successful in containing and/or remediating explosives contamination. Although research has demonstrated that plants are capable of converting TNT without the aid of microbes, very little TNT buildup has been discovered in plant material. Therefore, the potential groundwater treatment method of plant-enhanced breakdown of TNT is by aquatic macrophytes or phytoremediation.

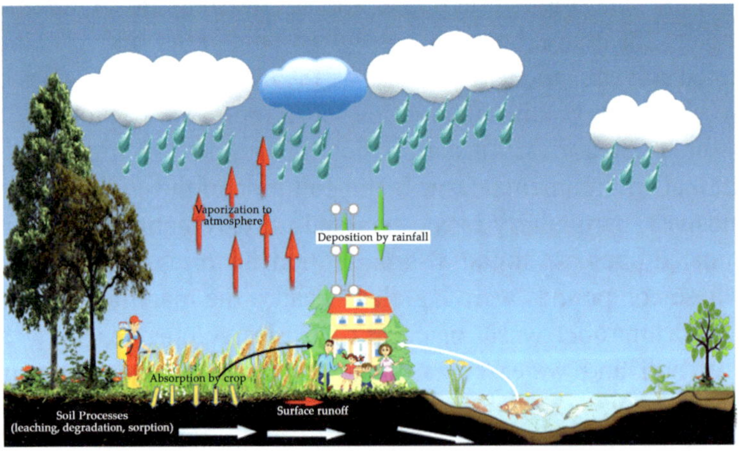

Figure 1.8. Pesticide behavior in the natural environment.

Source: *Tudi M, Li H, Li H, Wang L, Lyu J, Yang L, Tong S, Yu QJ, Ruan HD, Atabila A, et al. Exposure Routes and Health Risks Associated with Pesticide Application. Toxics. 2022; 10(6):335. https://doi.org/10.3390/toxics10060335*

Pesticides (Figure 1.8) have been widely dispersed throughout time due to heavy usage, mainly direct land application, most frequently in aquatic ecosystems. Due to the potential harm, they provide to both human health and the environment, the widespread use of pesticides has long been a source of public concern (Hussain et al, 2009). Pesticides can get into water bodies through point sources or diffuse sources. Diffuse source pesticide inputs come from pesticide field applications and can be caused by both agricultural and non-agricultural operations, having an impact on both surface and ground waters. These include surface and subsurface runoff, tile drain overflow, baseflow seepage, erosion, spray drift, deposition following volatilization, leaching through the soils, and infiltration via river banks and beds. The effluent from WWTPs, farmyard runoff that enters streams or the sewer system, and unintentional spills are the main point sources of pesticides and their treatment-resistant metabolites. The decrease of inputs from diffuse sources and point sources, as well as the degradation of contaminated areas, are all parts of the mitigation of pesticide contamination inputs into water bodies. Excavation and/or chemical oxidation procedures such as photocatalysis, ozonation, iron-catalyzed Fenton's reaction, as well as heat treatments, are traditional methods of remediating pesticide-contaminated soil i.e., low-temperature thermal desorption, incineration. In recent years, phytotechnologies have been utilized more frequently to remove the most enduring pesticides. There are still issues, such as the potential phytotoxicity of some substances such as herbicides that were created to obliterate plant matter (Watanabe, 2011).

Phytodegradation, rhizodegradation, and phytovolatilization have been frequently engaged in pesticide phytoremediation. Their versatility in function, which increases the potential for a variety of pesticide transport and degradation mechanisms, is often credited with their effectiveness. According to several studies, CWS has primarily been used to treat wastewater from point sources, such as WWTP. However, field studies have demonstrated that built wetlands and planted drainage ditches are also extremely successful in lowering pesticide concentrations and toxicity in non-point-derived wastewaters and runoff events. Therefore, it was suggested that vegetated agricultural water bodies serve as effective buffer zones for the defence of more delicate receiving waters from agricultural runoffs. Pharmaceuticals have recently attracted a lot of attention from the international scientific community among the various classes of organic xenobiotics, even though they have been poisoning numerous water bodies for a long time. The absence, until recently, of suitably sensitive methods of analysis for these low concentration ranges, coupled with the low concentrations of these compounds in the environment, has been the primary driver of the recent interest in the environmental issues posed by these compounds.

1.7 METABOLIC PROCESSES

Through metabolic excretion and manufacturing processes, chemicals used in human and veterinary medicine are continuously entering the environment.

This is primarily owing to the inappropriate disposal of unwanted or expired medications. Some of these pharmaceutical residues are released directly into the environment without undergoing the necessary treatment, but even those that do receive the necessary treatment and disposal in WWTPs are frequently not eliminated completely by the traditional wastewater treatment procedures (Di Gregorio et al, 2015). Human medicines are a diverse class of bioactive substances with a great deal of variation in their chemical structures, roles, behaviors, and activities. The majority of them are polar compounds since they are meant to be delivered inside organisms through an aqueous medium, and their creation and application are targeted at certain biological effects. Chemical compounds typically have molecular weights between 200 and 1000 Da.

Figure 1.9. Point sources of wastewater.

Source: *https://upload.wikimedia.org/wikipedia/commons/b/b7/Common_Point_Source_Discharges_-_EPA_2010.png*

Despite medications being used in greater numbers around the world, there is still a lack of information on their overall usage. Human pharmaceutical use and consumption may fluctuate significantly from one nation to the next because of variations in disease prevalence, treatment practices, and alternatives, or just for market considerations. For instance, while some drugs are available exclusively with a prescription in some countries, others sell them over-the-counter.

Point sources of wastewater (Figure 1.9) are the main entry points for human medications, their metabolites, and transformation products into the environment. Pharmaceuticals can enter the sewage system by excretion of unmetabolized substances and their metabolites or through the disposal of extra drugs through the sewage system. These substances typically evade effective removal by the traditional wastewater treatment methods in WWTPs.

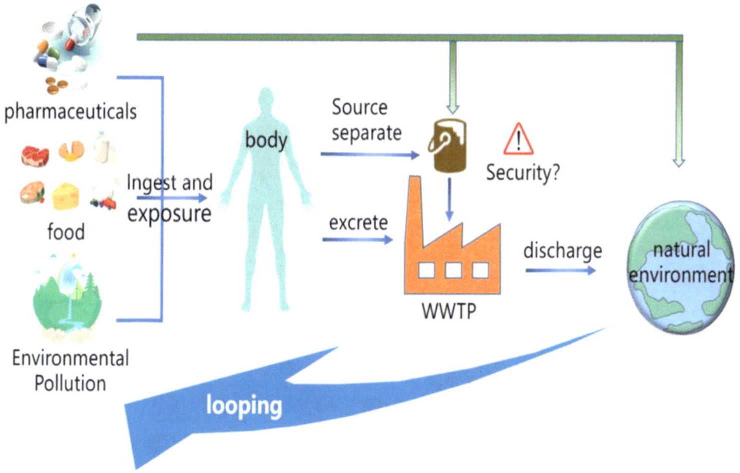

Figure 1.10. Pharmaceuticals in water.

Source: *Li X, Wang B, Liu F, Yu G. Occurrence and Removal of Pharmaceutical Contaminants in Urine: A Review. Water. 2023; 15(8):1517. https://doi.org/10.3390/w15081517*

1.8 PHARMACEUTICALS

Pharmaceuticals (Figure 1.10) are potentially prone to the same types of movement, transfer, and transformation/degradation processes as other organic pollutants after they are discharged into the environment via the discharge of treated wastewater. As a result, the way that medications interact with soil, surface water, and groundwater is very complicated. Pharmaceuticals may undergo sorption, hydrolysis, biological transformation/degradation, redox reactions, photodegradation, volatilization, and precipitation/dissolution during their transport and transformation processes in the aquatic environment. Pharmaceutical presence and mobility in aquatic ecosystems are influenced by these environmental processes, which are ongoing (Kotyza et al, 2010). Drug behavior along any of these channels for partitioning, degradation, or transformation may help to lower or eliminate environmental drug concentrations, reducing the likelihood that these medicines will have an adverse effect on aquatic life and human health. The most likely pharmaceutical compounds to reach steady-state levels in the environment and be detected in surface and ground waters as well as drinking water supplies are those that are marketed in large quantities, are water-soluble or slightly soluble, and are resistant to degradation through biological or chemical processes. Pharmaceutical chemicals, in general, have passed through the human digestive tract and perhaps a conventional WWTP, which is a significant distinction between them and other environmental POPs such as pesticides, PCBs, PAHs, explosives, and chlorinated solvents. Many medications will enter the aquatic environment in a transformed form as a result of this pre-exposure to biochemical metabolism, and those that are unmodified will, on the other hand, share a resistance to biotic transformation.

Given that many medications are water soluble, hydrolysis and photodegradation are the abiotic processes most likely to change them and remove them more permanently from the aquatic environment. Hydrolysis reactions, however, presumably play a less significant role in the aquatic fate of many medications that reach the environment given that they travel through the digestive system and have a relatively long residence time in aqueous settings within the WWTPs. On the other hand, for medicines with absorbances in the 290–800 nm range, direct photodegradation by sunlight may be a significant elimination mechanism. In any event, a number of variables related to both the pharmaceutical qualities and the environmental conditions regulate the amount of the various abiotic and biotic activities that may potentially have an impact on the immediate behavior and long-term fate of a pharmaceutical in the environment. This fate is influenced by the environment in addition to the qualities of the chemical. Temperature, sunshine, pH, the amount of organic matter in soils and sediments, and redox conditions are only a few of these variables. Pharmaceuticals often exhibit physicochemical characteristics that make them resistant to degradation and transformation under normal circumstances. As a result, they often have the potential to enter the environment. However, despite the low quantities at which they are shown to occur, little is known about the potential human or ecological dangers that can result from cumulative exposure to the cocktail of medicines and metabolites present in the various environmental compartments (Carvalho, 2021). Human medications are made to target particular metabolic and molecular pathways, but they may also have unintended effects on animal counterparts that have equivalent or identical target organs, tissues, cells, or macromolecules. Dissimilar pharmacological modes of action may arise even in animals with drug receptors that are absent or different from those present. It is crucial to understand that many medications have unclear precise mechanisms of action, and frequently not just one but several such modes of action take place. Thus, it is challenging to evaluate the ecotoxicity of the majority of medications. In addition, the metabolites and degradation by-products of medicines are of concern since many of them have toxicity levels that are often on par with or higher than those of the parent substances. The majority of the material now available on the ecotoxicological effects of human medications focuses on the short-term acute toxicity assessed through experiments with an emphasis on aquatic organisms. Most medications found in the environment have acute toxicity values in the mg/l dosage range, although levels in surface water are reported to be at least three orders of magnitude lower (Banerjee et al, , 2022).

1.9 TOXICITY

On the other hand, because there is typically a paucity of toxicity data, it is more challenging to determine whether long-term chronic impacts have any environmental importance. The presence of pharmaceuticals and related substances in the environment does, however, have some primary effects that can be identified, such as cumulative effects, endocrine disruption, the emergence of antibiotic-resistant bacteria, and genotoxic effects. When thinking about the significance of medications on the environment, persistence is just as important as toxicity. Many medications are not lipophilic, unlike POPs like insecticides, and do not bioaccumulate in the environment (Malik, Pirzadah, and Hakeem, 2022). Due to their ongoing entry into the environment, some of those substances are, however, considered pseudo-persistent pollutants. These chemicals regularly enter the environment, even though they don't have a long half-life, which exposes the aquatic ecology over time. Specialists have often brought up the problem of emerging pollutants like pharmaceuticals and the need to regulate water quality standards for this kind of pollution. Environmental agencies around the world are becoming more cognizant of this issue, designating these new substances as priority pollutants and mandating new environmental risk assessments be conducted as part of the procedure for approving new substances for use in the general public. Given this situation, it is expected that wastewater treatment standards would tighten in the years to come in terms of the limiting amounts of many of these compounds in the effluents from WWTPs. Many of the current conventional WWTPs will need to be modified or reformed in the upcoming years in order to comply with these new regulations. As a result, there is a rising need for alternative wastewater treatment methods that are more effective at removing pharmaceuticals from water and have low operating and maintenance costs. The use of phytotechnologies, such as constructed wetlands systems, maybe a different, more affordable option for the removal of pharmaceuticals from wastewater. These systems have already demonstrated high efficacy in removing some pharmaceuticals.

1.10 PHYTOREMEDIATION

Since the early 1990s, a broad term known as phytoremediation has been used to describe a collection of technologies that use plants, along with the associated microorganisms, enzymes, and water consumption, to remove, immobilize, transform, or degrade pollutants, primarily of anthropogenic origin, from soil, sludge, sediments, water and wastewater, and even the atmosphere. In contrast to other remediation strategies such as those involving excavation or chemical *in situ* stabilization/conversion, phytoremediation is appealing because it is relatively inexpensive and aesthetically pleasing to the public. Both *in-situ* and *ex-situ* phytoremediation techniques are available. Recent initiatives have concentrated on speeding up the decomposition of organic contaminants, frequently in collaboration with root rhizosphere bacteria, or removing dangerous heavy metals from soils or water. It has been determined that plants can lower the availability of contaminants in different environmental compartments through a number of processes (Sharma, 2021). These include hydraulic control, phytostabilization, phytoextraction, phytovolatilization, phytodegradation, rhizodegradation, and rhizofiltration. Pollutants accumulated in stems and leaves are harvested together with accumulating plants and removed from the site through the process of phytoextraction/phytoaccumulation, which involves the removal of pollutants by plant roots and subsequent transfer to aerial plant parts. Through the activity of internal or secreted enzymes, phytodegradation involves converting organic contaminants into molecules with lower toxicity. Soluble contaminants are absorbed with water by the roots, moved to the leaves, and then volatilized into the atmosphere through the stomata during the process of phytovolatilization. Pollutant transpiration quantities are typically minimal and proportionate to water flow. Rhizodegradation is the process by which organic pollutants are broken down by microbial enzyme activity. The kinds of plants that grow in contaminated areas affect the quantity, diversity, and activity of microbial populations (Ouvrard, Leglize, and Morel, 2014). Rhizofiltration is a method of retaining pollutants that either includes adsorption or absorption by plant roots, as a result, these processes typically call for enormous root surface areas. Plants growing on polluted sites also stabilize the soil and can serve as a ground cover, reducing wind and water erosion and direct contact of the pollutants with animals. Through the phytostabilization process, accumulation by plant roots or precipitation in the soil by root exudates immobilizes and reduces the availability of soil pollutants. Containment of pollutants within a site can also be achieved by limiting the spread of a contaminant. However, phytoremediation technology has its limitations and is not appropriate or effective for all sites and conditions.

Extremely high levels of pollutants may prevent plants from growing or even kill them. Therefore, phytoremediation is more efficient or restricted to pollutant concentrations below that. The size of the rooting zone limits the effective depth of phytoremediation, which is restricted to sites with shallow contamination. For phytoremediation to be effective, contaminants must be within reach of plant roots. Contaminants may accumulate in plant tissues as a result of phytoextraction. Environmental concerns include the possibility of ecological exposure from feeding animals such polluted plants and the potential effects on the food chain. Biomass that has been polluted may need to be harvested. The removal of pollutants from the subsurface through phytovolatilization may enhance airborne exposure (Chaney, Broadhurst, and Centofanti, 2010). In general, the fate of contaminants is frequently unknown, which may provide significant problems with the potential for some pollutants to spread across the environment. When non-native species are chosen for phytoremediation, unanticipated effects on the environment may result from their introduction. When compared to other treatment approaches, phytoremediation may require more time to reach remediation goals. For instance, it can take several growing seasons for a tree to establish itself and for pollutant levels to drop. Depending on the environment, phytoremediation may have seasonal success. Depending on the climate, it may or may not be effective. In order to prevent contamination from seeping downward or from spreading horizontally, trees can be employed as a hydraulic control barrier to direct water upward in the root zone. Additionally, plants can be utilized to stabilize contaminants in soil by obstructing their erosive, leaching, or runoff properties, or by transforming them into less bioavailable forms, such as through precipitation. The aerial plant components can then be harvested and used for non-food reasons (wood or cardboard) or they can be dumped in a landfill, among other uses, for phytoextraction of pollutants. When it comes to precious metals, recycling the accumulated element is an option; this process is known as phytomining (Liuwei Wang et al., 2020).

A specific pollutant removal mechanism may predominate in some phytotechnologies, although there are other potential mechanisms, whicht are not mutually exclusive. On the other hand, in the case of some phytotechnologies, many mechanisms can take place concurrently or cooperatively. For example, in a more sophisticated application of phytotechnology, such as a constructed wetlands system, rhizodegradation and phytostabilization, phytoextraction, phytodegradation, and phytovolatilization can all make varying degrees of contributions to the system's overall effectiveness in removing pollutants.

Constructed wetlands systems (CWS) are engineered systems created and built to harness natural processes including microbial assemblages, wetland vegetation, and soil to aid in wastewater treatment. The purification of wastewater attained in a CWS is the result of the coordinated activity of all these elements through a range of chemical, physical, and biological processes. Many of the same processes that occur in natural wetlands are utilized by these systems, but they do so in a more regulated setting. According to the type of water flow regime, CWS can be divided into several different types. They are free water surface (FWS) constructed wetlands, vertical subsurface flow (VSSF), and horizontal subsurface flow (HSSF). CWSs have been employed to treat a variety of wastewater from municipal sectors, and industrial waste. CWS have historically been utilized mostly as alternatives to or additions to the traditional treatment for household wastewater in small towns.

CWSs have therefore been used mostly in the removal of bulk wastewater contaminants such as pathogens, organic debris, excess nutrients, and suspended particles. More recently, CWS applications for addressing more specialized contaminants, like organic xenobiotics, have been attracting additional interest and becoming the focus of more investigations.

1.11 CONSTRUCTED WETLANDS SYSTEMS (CWS)

In contrast to the treatment of bulk contaminants, specialized use of CWS for the removal of certain organic chemicals or groups of compounds has been developing. Wastewater contaminated with organic xenobiotics from the petroleum industry, food processing industry, pesticide-contaminated agricultural runoff, landfill leachates, and waters containing surfactants, solvents, and mineral oils have all been the subject of significant research. The precise routes for the elimination of organic xenobiotics in CWS are not well understood. Given the enormous range of chemical properties of these chemicals, which, despite sharing the label "xenobiotics," are made up of chemical families that are distantly unrelated to one another, it is not surprising that they are eliminated via many different methods. This subset of xenobiotics is equally diverse in terms of its chemical features; therefore, it is possible to draw the same conclusion about pharmaceuticals in particular. The information that has been gathered is now starting to be implemented, even if most of the design of CWS in the past was done with little understanding of the functions played by each component and how its function could be enhanced. There is currently being introduced a far wider range of plant species, matrix materials, and artificial wetlands designs. The abiotic removal of pollutants in CWS involves a wide variety of physical and chemical methods. Sorption, which causes the pollutants to be temporarily retained or permanently immobilized, is the most significant abiotic removal activity taking place in CWS at the surface of plant roots and solid medium. The incidence of sorption processes will be strongly influenced by the kind of materials used to build the support matrix. The ability of a solid matrix to adsorb pollutants depends on its chemical properties. However, retention in abiotic components also depends on the wastewater's dissolved organic matter content, pH, and electrolyte composition, as well as the pollutant itself. Additionally, for the good and uniform contact of the wastewater with the CWS media and, consequently, for the efficiency of the system, a good hydraulic conductivity of the support matrix is essential. This prevents the occurrence of overland flows and preferential channeling.

Depending on their unique characteristics, other typical abiotic processes including hydrolysis, photodegradation, redox reactions, and volatilization can also contribute to the removal of some specific types of compounds to varying degrees. Abiotic processes, except for sorption, are typically not the primary removal methods for the majority of organic chemicals, such as pharmaceuticals, because the CWS characteristics of the compounds make them unsuitable. For instance, photodegradation in FWS systems can only be significant when the compounds are photosensitive and plant density is not excessively high. Since the water level in SSF systems is below the surface of the solid matrix, there is very little exposure of the pollutants to sunlight in these types of systems, and photodegradation does not occur to a significant extent. For compounds with little volatility, like the majority of medications, the volatilization process is likewise only marginally significant. Since organic compounds have already been subjected to these types of processes in the secondary treatment stage, they typically do not experience noticeable hydrolysis when CWSs are used as a tertiary treatment stage after conventional secondary treatment in a WWTP. Therefore, the organic xenobiotics found in CWS influences are either the end products of chemicals that did not resist hydrolysis or they are those substances that have resisted such processes. Through their ability to absorb organic xenobiotics in their roots and even uptake them and sequester/transform them in their tissues, plants are being given an increasing amount of credit in CWS for the removal of poorly or non-biodegradable organic xenobiotics. In fact, the elimination of inorganic contaminants through direct uptake by plants is a method that is widely accepted. Plant absorption is one of the most important biotic processes for some organic chemicals, especially for those with moderate hydrophobicity.

However, it is well acknowledged that the primary method for the removal of organic xenobiotics in wetlands is through the action of microbes. The influence of exudates released in the rhizosphere, which have the effect of stimulating the development and activity of microorganisms, does however allow plants to play a significant role in these microbial processes. Although microbes can also store and absorb substantial amounts of contaminants, it is their metabolic processes that are most important for the breakdown of organic chemicals by converting complicated molecules into simpler ones. This offers a crucial biological mechanism for the elimination of several chemical substances. However, the effectiveness and rate of the breakdown of organic molecules by microorganisms varies greatly depending on the type of compound. The characteristics of the biotic components (vegetation and microbes) have a significant impact on the CWS behavior as well. In CWS, microbial populations grow spontaneously and are subjected to the same conditions that influence their growth in WWTPs. However, by inoculating the CWS with strands that are more suitable for the system's goal, one can change the features of these microbial communities. The ability of microorganisms and plants to tolerate the more harmful contaminants and to either biodegrade or assimilate them are crucial traits. Other variables about the CWS design in the case of the vegetation, such as plant density and specimen arrangement such as how specimens of various species may be mixed when planted in the beds, all need to be taken into account and carefully planned as their influence may range from minor variations in the system's behavior to more pronounced impacts on the overall efficiency. Particularly, seasonal changes in the effectiveness of the system may be caused by cycles of physiology of select species in addition to variations in the climate, which in some situations may be minimized by using polycultures of vegetation. Although many organic contaminants are easily absorbed by plants, because they are xenobiotics, there are no particular transporters for these substances in the membranes of plants. Diffusion is the method used to transport them into and within plant tissues as a result. The water potential gradient that transpiration creates throughout the plant, which is influenced by the features of the plant and the CWS environmental circumstances, is what drives the flux. The concentration of the chemicals and physicochemical characteristics such as water solubility, the n-octanol/water partition ratio or coefficient or log Kow, and the negative base -10 log of the acid equilibrium constant (Ka) (pKa) are highly correlated with their translocation (Ouyang, 2002).

There might be an ideal hydrophobicity where the organic chemical has a propensity to bind to the lipid bilayer of the membrane but not too firmly, allowing for continued transit. For moderately hydrophobic organic pollutants, direct uptake by roots is typically an effective removal method. The lipophilic root and cell walls are typically so tightly linked to hydrophobic substances that they are unable to easily enter and be translocated in the plant. On the other hand, substances that are quite water soluble are neither successfully transported through the lipid bilayer of plant membranes nor properly sorbed to roots. However, recent research suggests that a log Kow value of a molecule may not be the only element influencing how likely it is to be taken up. Some compounds have even been demonstrated to be able to enter plant membranes while having a low log Kow. Another aspect that may affect a plant's ability to extract a substance from water is its anatomy and root system, as well as the initial pollutant concentration. Due to their prolonged residence durations in the root zone, extremely hydrophobic compounds are also suitable for phytostabilization and/or rhizosphere bioremediation. As the amount of substance absorbed by the plant determines the overall amount of xenobiotics available for these processes, uptake has the major influence over translocation, metabolism, and phytotoxic activity.

By making the substance less or more active, metabolism affects both xenobiotic absorption and their phytotoxic effect. The modeling of organic chemical absorption and transport in plants has made considerable use of the TSCF (Ibrahim and El Afandi, 2020). Transpiration Stream Concentration Factor or TSCF. The TSCF is estimated by computing the amount of the compound accumulating in the shoot per volume of water translocated.

TSCF= (Concentration in transpiration stream)/(Concentration in external solution)

There is no proof of active absorption (TSCF > 1) of xenobiotic organic chemicals, maybe except a few hormone-like substances such as phenoxy acid herbicides. When a chemical's uptake is not precisely proportional to water uptake (TSCF = 1), it is said to be excluded even though the mechanism of uptake is still believed to be a passive process. However, elements including the solubility of a xenobiotic in xylem sap and membrane permeability may restrict the extent or kinetics of passive uptake. In-plant contaminant sorption and quick metabolism may also lower xylem concentrations and prevent recorded TSCF levels from reaching one.

1.12 KINETICS OF XENOBIOTICS

According to several empirical relationships between the TSCF and the log Kow of the xenobiotic reported for organic chemicals, these values are related by recognizable bell-shaped Gaussian curves. These reveal lipophilicity that is optimal for absorption and translocation and imply that chemicals that are either extremely polar (log Kow 0.5) or highly lipophilic (log Kow > 3.5) will not be absorbed by plants. These predictive schemes, however, may not be applicable for some non-ionizable, highly water-soluble organics, according to laboratory and field experiments with xenobiotics like 1,4-dioxane, methyl tert-butyl ether (MTBE), sulfolane, and diisopropanolamine. Organic xenobiotics that enter plant cells are subject to metabolic changes that might cause them to partially or completely degrade, or they could convert them into less harmful chemicals and become bound in plant tissues. The term "green liver" is often used to describe plant systems because the metabolism of foreign substances is usually thought to be a "detoxification" process that is comparable to the metabolism of xenobiotic compounds in humans (Reichenauer and Germida, 2008). In higher plants, there are three probable stages in the metabolic transformation of organic substances: Phase I - Functionalization is the conversion or activation of lipophilic xenobiotic compounds; during this phase, the molecules of the hydrophobic compound undergo enzymatic changes to acquire a hydrophilic functional group such as a hydroxyl, amine, carboxyl, or sulfhydryl. These processes increase the compound's polarity and water solubility, which also increases the affinity of the compound for enzymes that catalyze additional transformations by the addition or exposure of the appropriate functional groups. Some xenobiotics may undergo oxidative breakdown at low concentrations intoCO_2 and common cell metabolites. Following this route, a plant cell detoxifies the substance and absorbs the resultant carbon atoms for use in the cell. High concentrations prevent complete detoxification and expose the toxin to conjugation.

1.13 ENZYMES INVOLVED IN XENOBIOTIC METABOLISM

Several classes of enzymes are known to have a significant role during this phase. In plants, cytochrome P_{450} monooxygenase plays a major role in the oxidative metabolism of xenobiotics. This enzyme is essential for bioactivation of xenobiotics into chemically reactive electrophilic molecules, which are then converted into conjugates during Phase II. Another significant category of enzymes that aid in the conversion of some xenobiotics is peroxidases. Some substances may also be subject to oxidation by peroxygenases (Morant et al, 2003). Laccase is required to disassemble aromatic ring structures, while Nitroreductase is required to break down nitroaromatics (Figure 1.11).

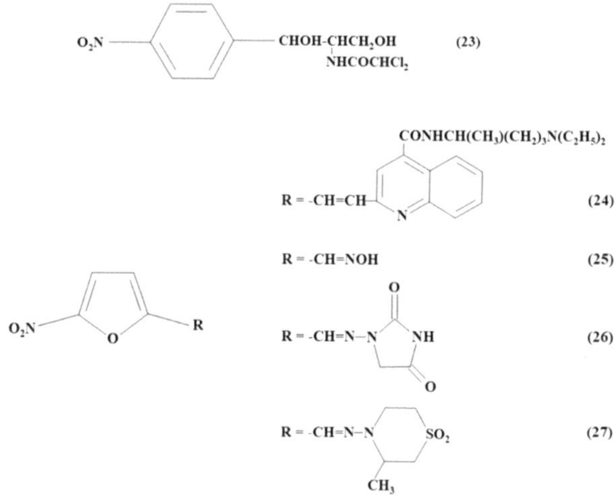

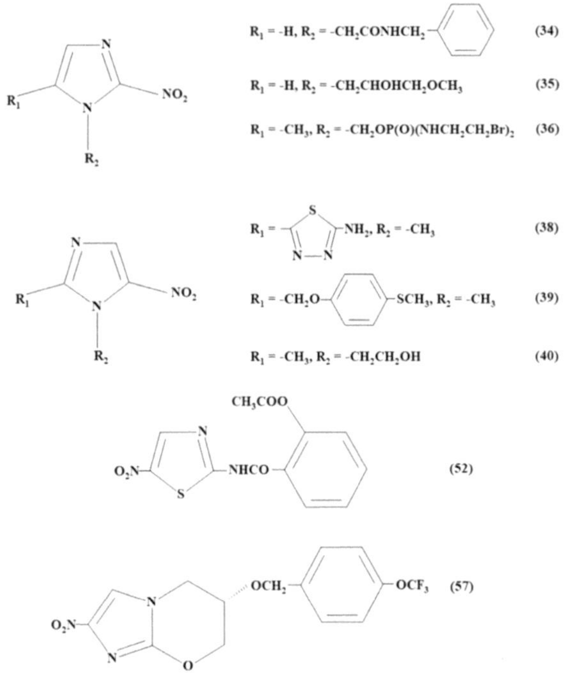

Figure 1.11 Formulas of nitroaromatic antibacterial and antiparasitic agents: chloramphenicol (**23**), chinifur (**24**), nifuroxime (**25**), nitrofurantoin (**26**), nifurtimox (**27**), benznidazole (**34**), misonidazole (**35**), TH-302 (**36**), megazol (**38**), fexinidazole (**39**), metronidazole (**40**), nitazoxanide (**52**), and PA-824 (**57**).

Source: *Čėnas N, Nemeikaitė-Čėnienė A, Kosychova L. Single- and Two-Electron Reduction of Nitroaromatic Compounds by Flavoenzymes: Mechanisms and Implications for Cytotoxicity. International Journal of Molecular Sciences. 2021; 22(16):8534. https://doi.org/10.3390/ijms22168534*

Phase I reactions alter the molecule to prepare it for Phase II and Phase III reactions, which further detoxify the xenobiotic. Phase I reactions are the first step required to finally make a xenobiotic less toxic. The chemical can, however, be employed immediately for Phase II metabolism without going through Phase I if it already has a functional group suited for that phase of metabolism. Conjugation occurs in Phase II and involves the conjugation of Phase I xenobiotic metabolites to endogenous molecules like proteins, peptides, amino acids, organic acids, monosaccharides, oligosaccharides, and polysaccharides, pectins, lignin, etc. Phase II final products are less harmful than either the original compounds or their Phase I derivatives. Phase II is linked to enzymes such as glutathione-S-transferases, glucosyl transferases, and N-malonyl transferases. The cytosol is where Phase I metabolites are conjugated, however, too much of these substances might be detrimental. Phase III - Compartmentalization entails the compartmentalization of modified xenobiotics in vacuoles or their binding to elements of cell walls like lignin or hemicellulose. The conjugates are transported to locations where they may not interfere with cellular metabolism during this phase. Soluble conjugates coupled with peptides, sugars, amino acids, etc. are accumulated in vacuoles, whereas insoluble conjugates coupled with pectin, lignin, xylan, and other polysaccharides are removed from the cell (Morikawa, Takahashi, and Kawamura, 2003).

Plants only experience phase III reactions since they do not excrete xenobiotics as animals do. Therefore, plants must eliminate xenobiotic from their systems. The primary enzymes involved in this phase are ATP-driven vacuolar transporters, and further conjugate processing may happen in the vacuolar matrix. However, this area of xenobiotic fate in plants is little understood, particularly with regard to the identity of the sequestered products and any potential fate in herbivores that consume plants. It is considered that Phase III compounds are no longer harmful. Every plant has defence mechanisms to guard against the damaging effects of trace amounts of foreign substances. Inferred from the physiological, biochemical, and molecular responses of the plant, the relative rates of organic xenobiotic absorption, translocation, and metabolism often indicate whether or not these substances will trigger a plant response. Physiological reactions including tissue necrosis, chlorosis, and growth inhibition are typically simple to observe. The relative growth rate (RGR) of the plant or the concentration of photosynthetic pigments in plant tissues are two quantitative indicators that can be examined in order to measure the physiological harmful responses to xenobiotic stress. These two factors can change when plants are subjected to xenobiotics, which can occasionally compromise plant viability. One of the most crucial targets of stress in plants is the photosynthetic system. In fact, the majority of metabolic responses brought on by stress conditions have an impact on the plant's ability to maintain an effective light energy conversion. Plants exposed to stress have been observed to have altered contents of total chlorophyll, chlorophyll a, and chlorophyll b and carotenoids. In order to protect themselves against the production of free oxygen radicals, stressed plants typically increase the amount of carotenoid pigments in their tissues. It is common to see drops in both total chlorophyll and the ratio of chlorophyll to carotenoids. Different species have shown these alterations in photosynthetic pigments when exposed to trace metals and organic xenobiotics like herbicides.

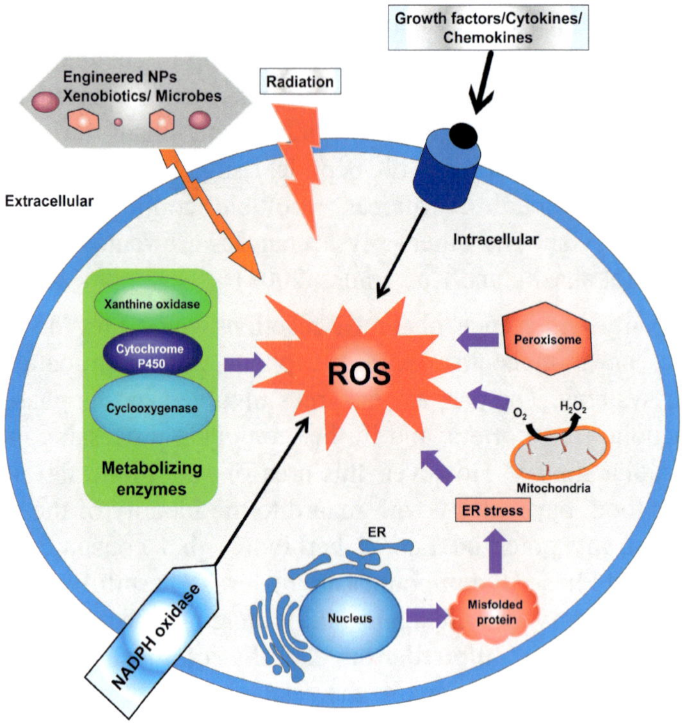

Figure 1.12. Reactive oxygen species.

Source: *Abdal Dayem A, Hossain MK, Lee SB, Kim K, Saha SK, Yang G-M, Choi HY, Cho S-G. The Role of Reactive Oxygen Species (ROS) in the Biological Activities of Metallic Nanoparticles. International Journal of Molecular Sciences. 2017; 18(1):120. https://doi.org/10.3390/ijms18010120*

1.14 ROS

Reactive oxygen species (ROS) (Figure 1.12) are produced as a result of biochemical changes brought on by the presence of organic xenobiotics. These chemical species are oxygen-based compounds that have been partially reduced (O_2) (Gupta, Huang, and Corpas, 2013). They typically result from either the transfer of one, two, or three electrons to oxygen to form superoxide radicals (O^{2-}), hydrogen peroxide (H_2O_2), or O_2^{3-} (which is broken down into water and hydroxyl radicals, OH) or from the excitation of the triplet O_2 to form singlet oxygen. Plants often manufacture these as by-products of several metabolic processes that take place in various cellular compartments, primarily in chloroplasts, mitochondria, and peroxisomes, including photosynthesis and respiration. Under physiological steady-state settings, various antioxidative defence elements that are frequently compartmentalized scavenge these compounds (Figure 1.13).

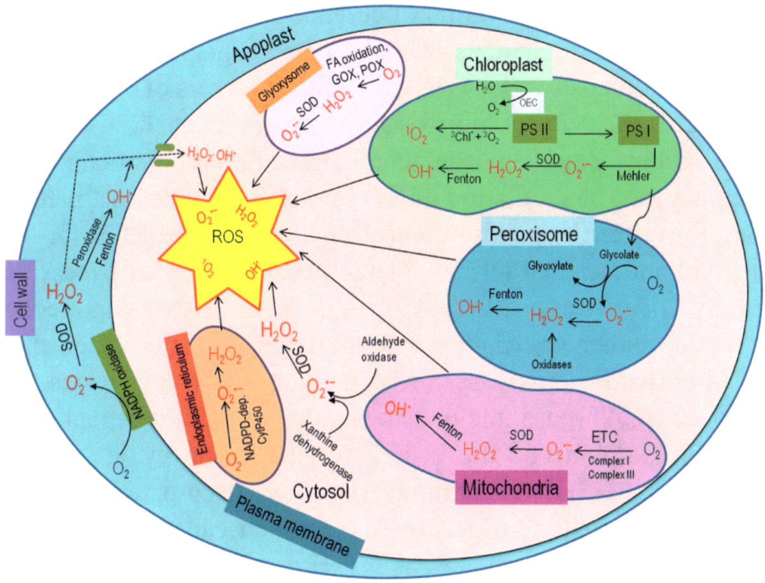

Figure 1.13. Reactive oxygen species-generation in plants.

Source: *Hasanuzzaman M, Bhuyan MHMB, Parvin K, Bhuiyan TF, Anee TI, Nahar K, Hossen MS, Zulfiqar F, Alam MM, Fujita M. Regulation of ROS Metabolism in Plants under Environmental Stress: A Review of Recent Experimental Evidence. International Journal of Molecular Sciences. 2020; 21(22):8695. https://doi.org/10.3390/ijms21228695*

Plants activate metabolic pathways to break down xenobiotics, which are foreign substances that produce a lot of ROS and can disturb the equilibrium between ROS generation and scavenging. These disruptions may cause intracellular ROS levels to spike quickly, endangering the viability of the cells. The quick synthesis of large amounts of ROS leads to an event known as an oxidative burst. Unlike atmospheric oxygen, ROS can oxidize a variety of cellular components without restriction. If they are not managed, ROS can harm biomolecules such as DNA and RNA damage, protein oxidation, and enzyme inhibition and eventually cause the oxidative death of the cell.

For these reasons, it is essential to precisely regulate and maintain the ROS levels within the cells within a specific range. Plant cells have evolved systems for detecting and removing excessive ROS. The ability of plants to handle excessive ROS generation is correlated with their tolerance to contaminants (Rizwan et al., 2018).

Despite these issues, the presence of ROS in the cells is necessary because they signal the activation of defence and stress response pathways. Thus, ROS can be thought of as secondary messengers in the stress response signal transduction pathway as well as cellular markers of stress. As a result, determining the activity of antioxidant enzymes has been used frequently to determine how much environmental stress different contaminants have caused in plants. The regulation of ROS involves two distinct mechanisms: one that allows for the precise control of low levels of ROS for signaling reasons, and another that allows for the detoxification of excessive ROS, particularly under stress. All plants include ROS detoxifying mechanisms, which can be either enzymatic or non-enzymatic. Superoxide dismutase (SOD), ascorbate peroxidase (APX), glutathione peroxidase (GPX), and catalase (CAT) are major enzymatic ROS scavengers in plants. Unlike CAT, APX needs the ascorbate-glutathione cycle, an ascorbate and glutathione (GSH) regeneration pathway. Ascorbate is converted to monodehydroascorbate (MDA) by APX, which can then be recycled by MDA reductase (MDAR) using NAD(P)H as a reducing agent. Ascorbate and dehydroascorbate can spontaneously dismutate from MDA (DHA). Dehydroascorbate reductase (DHAR), which is activated by the oxidation of GSH to oxidized glutathione (GSSG), mediates ascorbate regeneration (Arnao and Hernández-Ruiz, 2019).

1.15 PHENOLIC CHEMICALS

Phenolic chemicals, (Figure 1.14) which are common plant secondary metabolites and are involved in detoxification and defence against oxidative stress, may play a significant role in both enzymatic and non-enzymatic pathways. Numerous phenolic substances serve as substrates for antioxidant enzymes, including guaiacol peroxidase. Additionally, because of their ability to quench radicals due to their chemical structure, phenolic compounds have antioxidant characteristics on their own and are significant contributors to the non-enzymatic regulation of the oxidative burst (Harvey et al., 2002). In order to evaluate the oxidative stress brought on by the exposure to xenobiotics, a quantitative parameter that is complementary to the evaluation of antioxidant enzyme activities is the variation of the levels of phenolic compounds in plants.

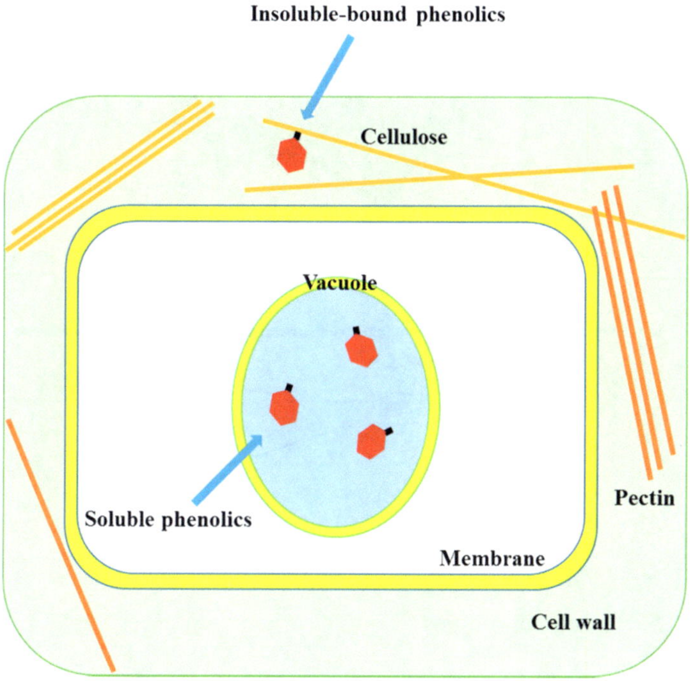

Figure 1.14. Phenolic chemicals in plant cells.

Source: *Shahidi F, Yeo J. Insoluble-Bound Phenolics in Food. Molecules. 2016; 21(9):1216. https://doi.org/10.3390/molecules21091216*

Numerous studies have been conducted on the toxic effects that particular classes of xenobiotic organic compounds can have on plants, including *Typha* spp. The oxidative stress brought on by pharmaceuticals is still poorly understood and characterized, which adds to the general lack of information needed to develop effective phytoremediation strategies for pharmaceutical contamination. Some of the more effective or conventionally utilized technologies are expensive, which prevents them from being employed in most circumstances. Plants and associated rhizosphere microorganisms have been used for a long time as a low-cost alternative to treat wastewater and clean up contaminated sites. This has been done through a variety of phytotechnologies, where the targets that need to be cleaned up and the mechanisms that work during the cleaning may differ greatly.

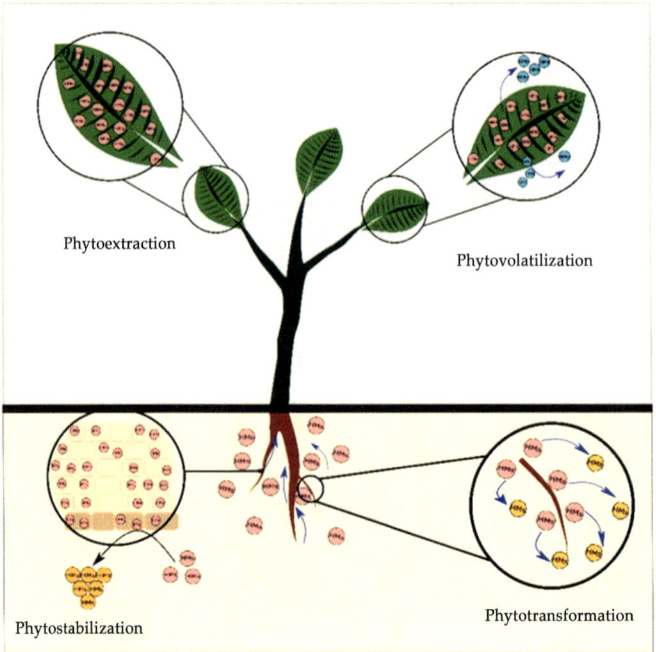

Figure 1.15. Phytotechnologies.

Source: *Raklami A, Meddich A, Oufdou K, Baslam M. Plants—Microorganisms-Based Bioremediation for Heavy Metal Cleanup: Recent Developments, Phytoremediation Techniques, Regulation Mechanisms, and Molecular Responses. International Journal of Molecular Sciences. 2022; 23(9):5031. https://doi.org/10.3390/ijms23095031*

1.16 PHYTOTECHNOLOGIES

Phytotechnologies (Figure 1.15) have numerous benefits, including aesthetically pleasing installations and generally positive public acceptance, but their applicability may occasionally be limited by the typically lengthy times needed to achieve remediation goals and the restriction to low concentrations of pollutants due to their potential toxicity to plants (Waigi et al, 2017). Despite some of its drawbacks, phytotechnologies have been employed to remove a variety of POPs, both from contaminated soils and water, with great effectiveness. The removal processes may involve a variety of mechanisms to varying degrees; some of these mechanisms may be abiotic in nature, but the most significant ones may be biotic, either as a result of direct plant action or as a result of degradation by microorganisms that have been stimulated by the plants. Constructed wetlands systems are one of the many forms of phytotechnologies that are becoming more and more popular. Through a range of physical, chemical, and biological mechanisms, the coordinated activity of plant species adapted to water-saturated conditions, microbes typical of these systems, and minerals of the support matrix, have demonstrated good capacities to deal with POPs. These systems are designed to optimize the processes for removing pollutants in a regulated setting. In order to pick components more easily and to optimize design and operating conditions, it is essential to have a thorough understanding of the mechanisms underlying the removal processes as well as the roles that each component plays. There is still a considerable amount of work in this area that can and must be done. Pharmaceutically active chemicals are a subgroup of POPs that are a member of a category of so-called emergent pollutants and have become particularly concerning in recent years. The research of CWS applications for the removal of this particular class of POP was prompted by the necessity to treat these unique compounds, which were created to elicit a biological response and may therefore have particularly adverse effects.

CHAPTER 2

Environmental Contaminants and Phytoremediation

CONTENTS

2.1 INTRODUCTION

Brownfields are defined as real estate where the existence or potential presence of a hazardous material, pollutant, or contaminant may hamper its growth, redevelopment, or reuse (Figure 2.1). The gas station is a typical brownfield location holding onto polluted land. In the 20th century, gas stations sprang up all across the U.S. as large oil firms overbuilt their networks (O'Connor et al., 2019). In order to solve the widespread contamination problems in soils and groundwater, phytotechnologies have the potential to answer the growing need for an inventive, environmentally friendly, and affordable solution. A gas station can provide enjoyment as a public green area or ecological services that benefit society after a complete restoration. Phytotechnology is about redesigning the landscapes of the post-industrial period using carefully selected plants, installation techniques, and innovative design approaches. This concept focuses on landscape architecture because it takes into account natural systems, takes into account various scales between the site and the region, stresses preventative measures, incorporates green infrastructure, and addresses the need to take into account cultural values. In phytotechnology, vegetation is used to contribute nutrients, porosity, and organic matter as well as to remediate, contain, or prevent toxins from entering soils, sediments, and groundwater.

Figure 2.1. Brownfields.

Source: *https://upload.wikimedia.org/wikipedia/commons/9/9e/Soilcontam.JPG*

2.2 BROWNFIELDS AND THEIR ISSUES

The advantages of converting urban brownfields into public green spaces have been extensively documented. A number of diverse factors contribute to these advantages, including visual preference, urban biodiversity, rising property values, and favorable community surveys. Without a doubt, brownfields have the inherent potential to become green spaces for the community or the environment, which can significantly improve the built environment of a city, remove contaminants, and revitalize urban areas that have been neglected in terms of social and environmental issues. Nevertheless, only a small portion of brownfield redevelopment projects are geared toward these purposes. The enormous cost involved with many brownfield restoration initiatives is one of the main causes of this. The intangible, qualitative, and long-term advantages of developing green spaces are frequently overshadowed by revitalization driven by the promise of housing opportunities and economic gains through jobs, tax revenues, and increased revenues because of residential and commercial redevelopment. The challenges involved in the transition from urban brownfields (Figure 2.2) to green spaces must be particularly addressed in order to affect future prosperity (Guidi Nissim and Labrecque, 2021).

Figure 2.2. Examples of brownfields in the Górnośląsko-Zagłębiowska Metropolis. (**A**) Coal Mine Kazimierz Juliusz; (**B**) Textile Works in Sosnowiec.

Source: *Pytel S, Sitek S, Chmielewska M, Zuzańska-Żyśko E, Runge A, Markiewicz-Patkowska J. Transformation Directions of Brownfields: The Case of the Górnośląsko-Zagłębiowska Metropolis. Sustainability. 2021; 13(4):2075. https://doi.org/10.3390/su13042075*

2.3 CONTAMINANTS IN GAS STATIONS

The majority of contaminants found at normal gas stations are organic compounds, which have different forms and are from petroleum sources (Figure 2.3). Through fuel spills, leaking underground storage tanks, and faulty fuel delivery, these chemicals are released into the environment. The substances contain a large number of hydrocarbon molecules that have negative effects on people, animals, and the environment when they come into touch with them. They frequently have chemical compositions that include single molecule chains that are more water soluble and thus more readily degradable.

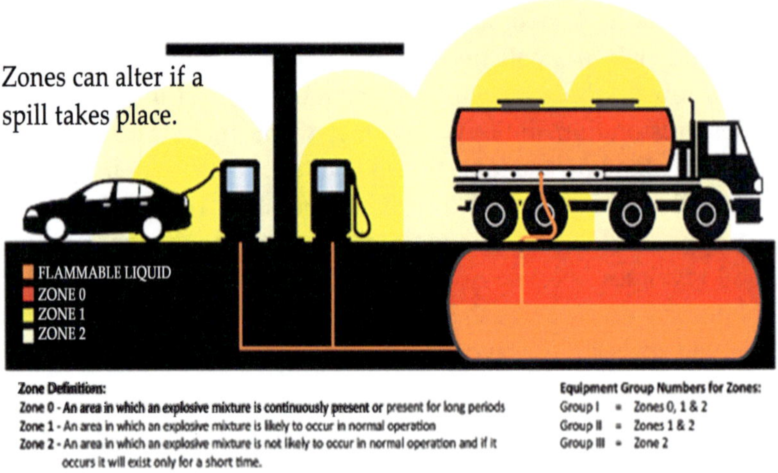

Figure 2.3. Hazardous zones in a gas station.

Source: *Ajman NN, Zainun NY, Sulaiman N, Khahro SH, Ghazali FEM, Ahmad MH. Environmental Impact Assessment (EIA) Using Geographical Information System (GIS): An Integrated Land Suitability Analysis of Filling Stations. Sustainability. 2021; 13(17):9859. https://doi.org/10.3390/su13179859*

Chlorinated solvents (Figure 2.4) are a different class of toxins that are common in petrol stations. They can be found in the thinners and degreasers used in auto maintenance and repairs.

Phytotechnologies can be used to target the group of chlorinated solvents and all lighter portions of petroleum hydrocarbons.

Organic pollutants might be degraded to the point that they are removed from the soil or groundwater as hazardous substances. Inorganics are substances that, in their purest form, cannot be broken down or degraded. There are less effective phytoremediation techniques for dealing with inorganic pollutants. Extraction and removal of plant matter is one method for addressing the problem of inorganic pollution. Inorganic compounds can be absorbed by plants through extraction, which retains them until the plant is ready to be harvested and moved somewhere for disposal.

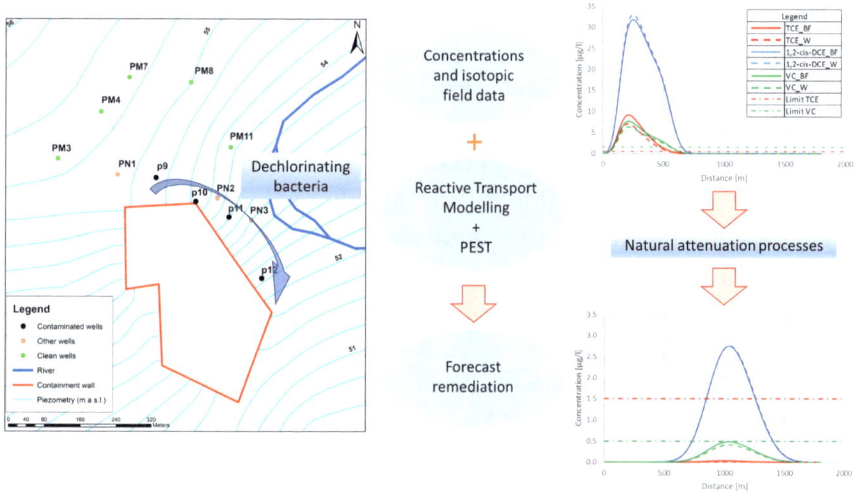

Figure 2.4. Chlorinated solvents.

Source: *Antelmi M, Mazzon P, Höhener P, Marchesi M, Alberti L. Evaluation of MNA in A Chlorinated Solvents-Contaminated Aquifer Using Reactive Transport Modeling Coupled with Isotopic Fractionation Analysis. Water. 2021; 13(21):2945. https://doi.org/10.3390/w13212945*

Applying phytoextraction in the field is difficult because it depends on a variety of elements, including the bioavailability of heavy metals, the characteristics of the soil, the speciation of the heavy metals, and the plant's capacity to absorb and accumulate metals in its above-ground components. However, it has been demonstrated that the technology of phytostabilization can be successfully used in the context of phytoremediation. It holds these contaminants in place with the help of flora, preventing them from flowing into important water sources or locations (Alkorta et al, 2010).

To satisfy design objectives and take into account functional needs, remediation methods can be spatially organized using phytotypologies. If necessary, several phytotypologies can be blended, or they can be incorporated with non-remediation planting techniques to incorporate aesthetic and ecoservice functions that are important to landscape architecture. By using a mix of techniques, a site with a complicated pattern of contamination can be handled to better encourage remediation while addressing other site-relevant concerns. Through the process of phytoirrigation, contaminated groundwater is extracted from underground pollution wells and pumped to the surface for reuse in irrigation. To avoid the release of toxins above the surface, where they could endanger wildlife, this should be applied by a drip irrigation system. The root zones of the crops may be exposed to toxins once they are pushed to the surface and released via the drip system. The plants must have a high evapotranspiration rate, fast-growing, and tolerance to petroleum in order to efficiently absorb water. This applies to phreatophytes with deep roots of up to 30 feet. The intercepting hedge is a technique for concentrating contaminated groundwater (Conger and Portier, 1997). Gas stations in more crowded urban areas can benefit greatly from this solution because it is appropriate for scenarios with a restricted quantity of space. To stop contaminated groundwater from seeping off-site, a single row of trees is often planted around the perimeter of a site and downstream of the source of the pollution. It is advised to use phreatophytes with extensive root systems. These hedges can act as a visual or aesthetic screen as well as a buffer for contaminants because they are used as site perimeters.

2.4 PHYTOTYPOLOGIES

Phytotypologies include the stormwater filter, degradation cover, green roof, and green wall in addition to the airflow buffer and interception hedge. Typologies can be distinguished in a useful way, yet the characteristics can also blend together.

Figure 2.5. Quercus rubra.

Source: *By No machine-readable author provided. Velela assumed (based on copyright claims). - No machine-readable source provided. Own work assumed (based on copyright claims)., Public Domain, https://commons.wikimedia.org/w/index.php?curid=398300*

A street airflow buffer, for instance, can serve as an intercepting hedge. Airflow Buffer is a line of large-leaved trees and a layer of grasses with varying-height shrubs that collect debris that is specific to traffic. Trees that can be possibly used are *Quercus rubra* (Figure 2.5) and *Fraxinus pennsylvanica* (Figure 2.6). A screen of strong, oil-tolerant plants like *Salix alaxensis* called an interception hedge stops polluted groundwater from leaking off-site (Sleegers and Hisle, 2018). In addition, the interception hedge acts as a barrier to airflow. The plants in the airflow buffer and pollutants that are rinsed off the street are both treated by vegetation. It acts as a barrier between the road and the defunct gas stations to shield the area from additional contamination.

(a) (b)

Figure 2.6. Fraxinus pennsylvanica.

Source: *Hauer RJ, Ball JJ, North E. Observation of External Wounding on Green Ash (Fraxinus pennsylvanica Marshall) Trees Associated with Tree Injection Systems. Forests. 2022; 13(11):1802. https://doi.org/10.3390/f13111802*

Metal environmental pollution, such as that caused by heavy metals (HMs), is a serious issue on a global scale, endangering soil ecosystems, surface and groundwater, as well as human health and food safety. The main human-caused factors causing this persistent pollution include inputs from burning fossil fuels, mining, ongoing industrial activity, municipal solid waste disposal, wastewater discharges or irrigation, and overuse of fertilizers and pesticides. A rise in polluted areas as a result of these human activities needs to be remedied to stop or lessen the transfer of contaminants into terrestrial, air, or aquatic habitats. Widespread concerns are raised by point and diffuse organic and inorganic pollution, and the intentional or unintentional release of these compounds into the environment could have a negative influence on human health.

2.5 SOIL CONTAMINATION

In the EU, soil contamination is a significant problem. In the EU, it was anticipated that 3.5 million sites were potentially contaminated, with 0.5 million of those sites being extremely contaminated and in need of treatment (Van Ginneken et al., 2007). These inorganic pollutants, such as HMs, which act as a final acceptor of toxic substances/trace elements, can limit plant development and growth and present a health risk to people and animals because some of them may be bioaccumulated and bio-magnified throughout the entire food chain. Environmental remediation is essential to reduce the danger of food chain contamination and other related health problems when the concentration of these pollutants is over a specified legal standard value in water, air, and soils. Therefore, effective soil clean-up methods are required to remediate soils that have been contaminated with heavy metals. To do this, various physical, chemical, and biological methods have been tried during the past few decades. Conventional methods of treating soils include digging up the soil and dumping it in landfills, washing the soil with water and solubilizing agents, erecting a physical cover, solidifying the soil by using stabilizing agents, vitrifying the soil at high temperatures, electrochemical separation, etc. These methods may quickly remediate the soil but irreparably harm the ecological environment. By utilizing plants and associated soil microorganisms to absorb, accumulate, immobilize, or biodegrade organic and inorganic pollutants present in various environmental matrices, phytoremediation is based on effective, affordable, and environmentally friendly rehabilitation strategies. A relatively new approach called phytoremediation is advocated as having sustainable costs and improving native microflora as well as physical, chemical, and biological qualities, improving soil health and fertility. Because harvested biomass can be utilized to produce renewable energies like biofuel production, phytoremediation also appears to enhance plant development and carbon sequestration.

Therefore, phytoremediation is a good choice for removing metals from contaminated soil. The low-cost phytoremediation techniques that are available for the remediation of organic contamination, such as phytodegradation and rhizodegradation, are not relevant to soils that have been contaminated with metals since metals are immutable. Additionally, the phytovolatilization method of phytoremediation is restricted to certain metals with volatile forms, such as mercury (Hg) and selenium (Se).

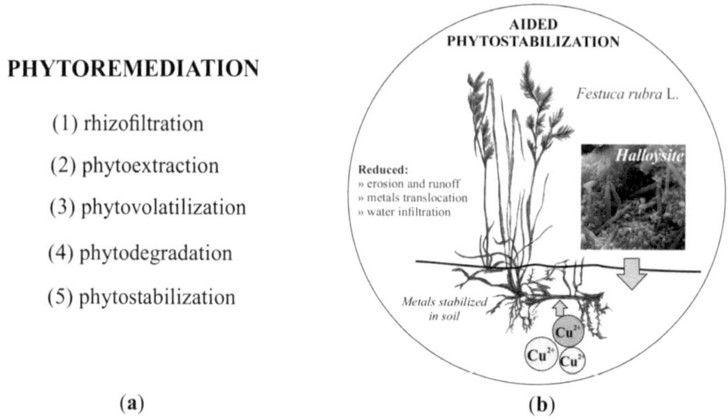

Figure 2.7. Phytostabilization- the concept of aided phytostabilization using halloysite and red fescue (Festuca rubra L.)

Source: *Radziemska M, Vaverková MD, Baryła A. Phytostabilization—Management Strategy for Stabilizing Trace Elements in Contaminated Soils. International Journal of Environmental Research and Public Health. 2017; 14(9):958. https://doi.org/10.3390/ijerph14090958*

Phytostabilization (Figure 2.7) and phytoextraction are the only feasible remediation strategies. Based on potential financial benefits, the goals of phytoremediation can be plant-based metal extraction or phytoextraction, risk reduction or phytostabilization, and sustainable soil management, in which phytoremediation gradually improves soil fertility to enable the growth of crops with increased economic value. It can be a successful method for the in-situ or ex-situ stabilization, eradication, or biodegradation of a wide variety of pollutants in the various environmental compartments, such as trace elements like HMs and non-essential metals, radionuclides, excess nutrients, salts, and recalcitrant organic pollutants like petroleum hydrocarbons (PHC), polycyclic aromatic hydrocarbons (PAHs), etc. The pollutant class, its bioavailability, particularly in soils, the physical and chemical properties of the matrix (soil, water, and wastewater), and the plant species play a role in how effectively phytoremediation works.

Figure 2.8. Metallophytes: examples tested as Ni-hyperaccumulator (**A**) A. utriculata and (**B**) N. caerulescens and non-hyperaccumulator (**C**) *A. montanum* and (**D**) *T. arvense,* photographs by S. Marsili (**A**,**C**), N. Mazzoni (**B**), G. Nicolella (**D**).

Source: *http Rosatto S, Mariotti M, Romeo S, Roccotiello E. Root and Shoot Response to Nickel in Hyperaccumulator and Non-Hyperaccumulator Species. Plants. 2021; 10(3):508. https://doi.org/10.3390/plants10030508*

The metallophytes (Figure 2.8) are the plants that are thought to be more effective at phytoremediation. These can thrive and reproduce in soil that has been contaminated with metals. Many known metallophytes, however, have low biomass and slow growth, which are disadvantages for phytoremediation technologies. Metal hyperaccumulators are another classification for some metallophytes (Awa and Hadibarata, 2020). These plants have higher concentrations of some metalloids than are typically necessary for healthy growth and development. Although they come from different families, hyperaccumulators have one thing in common: they can live on soils rich in metals and accumulate metals at levels much higher than those observed in the majority of species without experiencing phytotoxic consequences. Hyperaccumulators can be distinguished from related non-hyperaccumulating plants by three main characteristics: a significantly increased rate of metalloid uptake, a quicker root-to-shoot translocation, and an improved capacity for metal detoxification and sequestration in leaves (Yang et al, 2005).

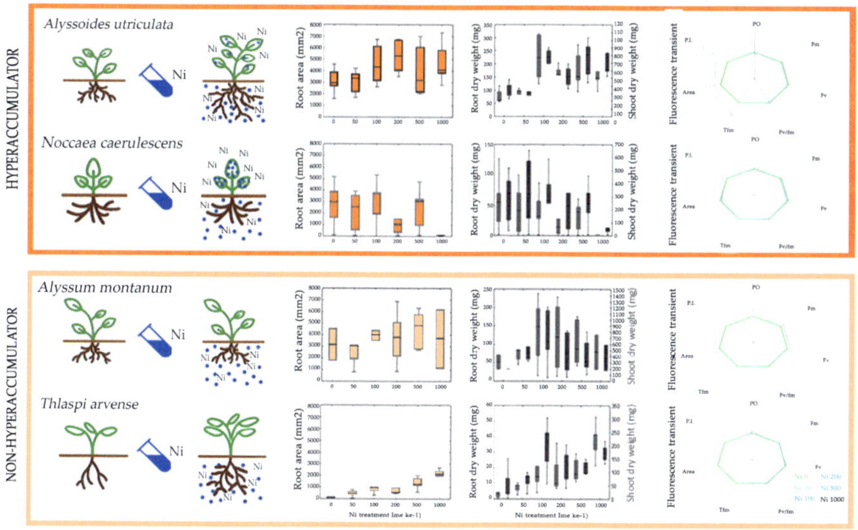

Figure 2.9. Hyperaccumulator plants

Source: Rosatto S, Mariotti M, Romeo S, Roccotiello E. Root and Shoot Response to Nickel in Hyperaccumulator and Non-Hyperaccumulator Species. Plants. 2021; 10(3):508. https://doi.org/10.3390/plants10030508

2.6 HYPERACCUMULATOR PLANTS

Hyperaccumulator plants (Figure 2.9) often can accumulate only one metal, whereas metal-polluted soils frequently contain a combination of metals. These plants can be exploited in phytoextraction applications. These plants store metals in the soil in their roots but limit the movement of metalloids within the plant. Such plants can be useful in phytostabilization procedures; however, they cannot be employed for phytoextraction. Non-hyperaccumulator plants with strong biomass production are some other plants employed for the phytoextraction of metals. These plants produce more surface biomass than hyperaccumulators while accumulating less metalloids than hyperaccumulators, eliminating comparable amounts of metals. Therefore, hyperaccumulators have a high potential detoxifying capability due to high uptake, an effective root-to-shoot transport system, and higher metal tolerance. High concentrations of metals, however, are toxic to the majority of plants and can harm cell metabolism, slow down plant growth, and limit metal phytoextraction. Enzymatic activity, protein structure, water balance, respiration and ATP content, photosynthesis, plant division, and morphogenesis are some of the physiological mechanisms that could be impacted. There are some drawbacks to phytoremediation, including the fact that many species cannot be grown in heavily polluted areas, the remediated area is only that close to the root, harvestable surface biomass is low, and naturally occurring hyperaccumulators develop slowly. Therefore, it may be more advantageous in this regard to use genetic engineering techniques to produce transgenic plants with better traits including high biomass production, higher metal accumulation, tolerance to metal toxicity, and good climate adaptation (Jesus et al, 2015).

In order to make these technologies more efficient, time-saving, and financially viable, additional research in the field of genetic engineering is required to enhance the phytoremediation capabilities of transgenic plants and to comprehend the mechanisms and effectiveness of phytoremediation techniques. As a result, new phytoremediation techniques for soils contaminated with metalloids are required. The use of soil additives and the employment of microorganisms related to the plants are strategies to boost the efficacy of phytoremediation.

Protozoa, algae, fungi, and bacteria are just a few of the diverse microbes that require the rhizospheric environment. These bacteria interact with plants in a variety of ways. Microbes can produce and detect the signaling molecules that cause microbial populations to assemble into a biofilm around the root surface and cause an equivalent reaction.

Numerous bacteria that are found in plants are advantageous because they can increase the bioavailability of nutrients and lessen the detrimental impact that metals have on plants. Rhizodegradation, biotransformation, and volatilization are three processes by which rhizospheric bacteria can biodegrade organic contaminants.

Numerous studies have shown how certain metalloids negatively affect the diversity of soil microbes and interact with plants in contaminated soils. The destruction of cellular membranes and organelles, denaturation of enzymes, and functional or molecular conformational disturbance are the principal mechanisms by which excessive concentrations of metal(loid) in soils might impact the growth, morphology, and metabolism of microorganisms (Kotrba et al, 2009).

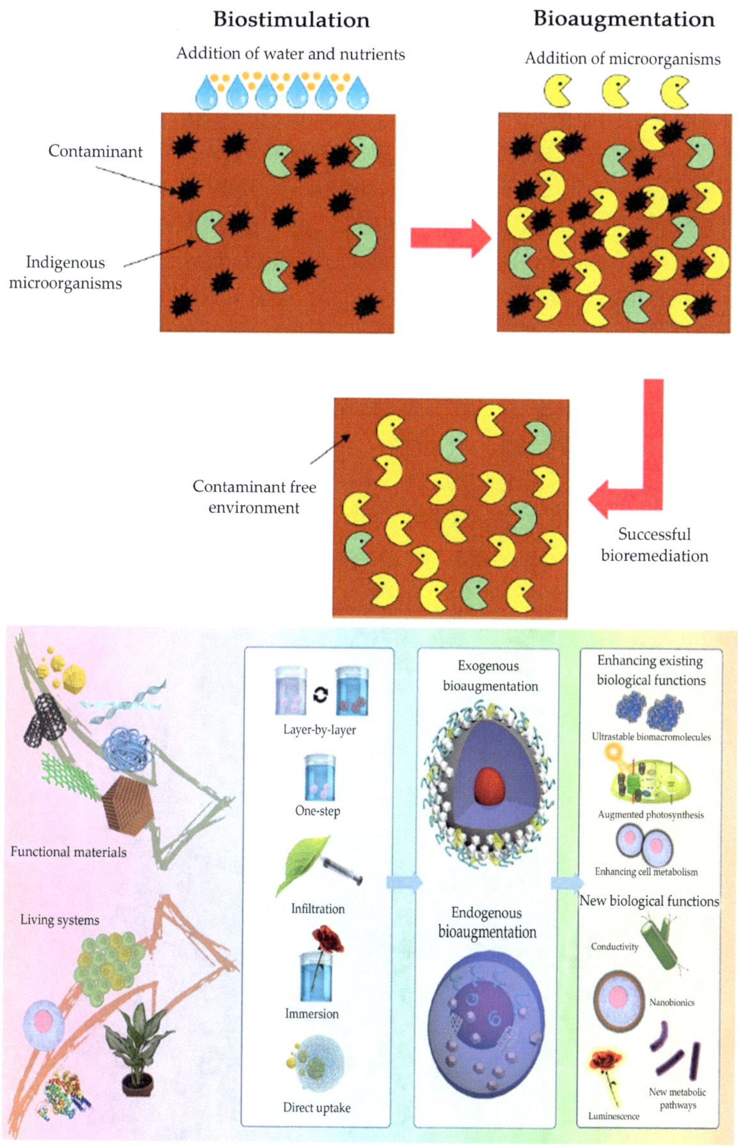

Figure 2.10. Bioaugmentation

Source: *Sales da Silva IG, Gomes de Almeida FC, Padilha da Rocha e Silva NM, Casazza AA, Converti A, Asfora Sarubbo L. Soil Bioremediation: Overview of Technologies and Trends. Energies. 2020; 13(18):4664. https://doi.org/10.3390/en13184664*

2.7 PHYTOREMEDIATION APPROACHES

As assisted phytoremediation approaches are to be employed to detoxify polluted soils on a broad scale, the inoculation of plants with specific and acclimated microorganisms (bioaugmentation) (Figure 2.10) has achieved prominence for phytoremediation of metal(loid) polluted soils (Agnello et al, 2016). The interaction of microorganisms and plant roots in the rhizosphere is thought to have a significant impact on plant development and survival. The biotechnological ability of microorganisms to withstand and/or directly remove metals from polluted media and their favorable impacts on plant growth may lead to economical and environmentally friendly methods for reclaiming polluted soils. By creating biosurfactants, organic ligands from the microbial breakdown of soil organic matter, and exudates such as metabolites, microbial siderophores, rhizosphere microorganisms can affect the bioavailability and uptake of HMs by plants. Therefore, metal speciation and transport in the environment are significantly influenced by microbial activity.

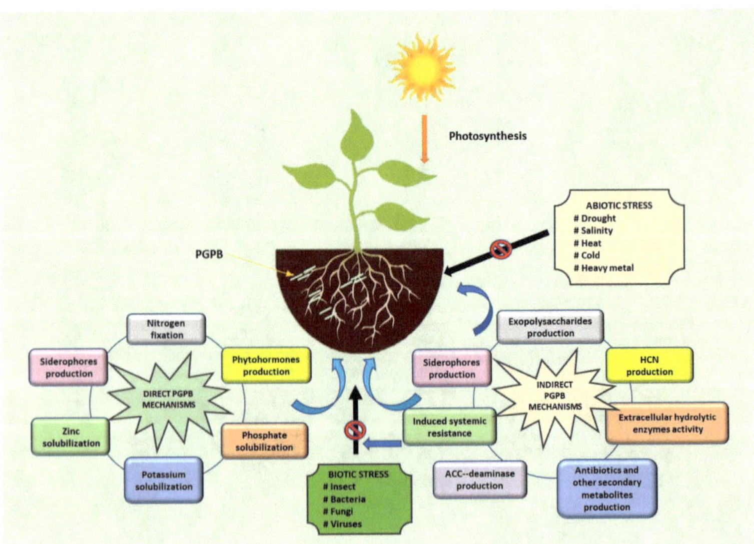

Figure 2.11. Plant growth-promoting bacteria and their functions in plants.

Source: *Ajijah N, Fiodor A, Pandey AK, Rana A, Pranaw K. Plant Growth-Promoting Bacteria (PGPB) with Biofilm-Forming Ability: A Multifaceted Agent for Sustainable Agriculture. Diversity. 2023; 15(1):112. https://doi.org/10.3390/d15010112*

Different organisms react to toxic ions in various ways, giving them a spectrum of metal tolerance. Microbe-assisted phytoremediation has become a viable method for cleaning up soil that requires improvement and causes little soil disturbance. By altering the rhizospheric environment, increasing biomass production and the bioavailability or stabilization of HMs, and lowering the corresponding toxicity, some species of microorganisms, such as biodegradative bacteria, plant growth-promoting bacteria (PGPB), (Figure 2.11) and filamentous fungi, appear to be helpful in phytoremediation (Ma, Rajkumar, Zhang, and Freitas, 2016b). They can therefore be applied to improve polluted soil. By altering the pH and oxidation-reduction potential of their soil environment, microorganisms can produce organic molecules that solubilize and/or stabilize HMs. Glomalin and other glycoproteins secreted by arbuscular mycorrhizal fungus (AMF) (Figure 2.12) enhance particle aggregation and aggregate stability against wind and water erosion, hence enhancing soil structure (Chen et al., 2022).

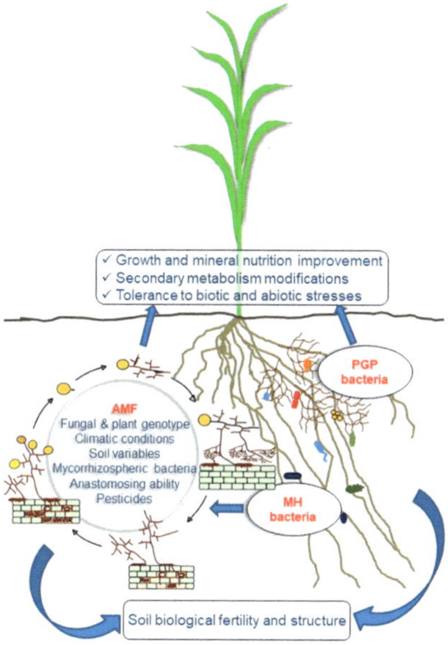

Figure 2.12. Arbuscular mycorrhizal fungi and bioremediation.

Source: *Giovannini L, Palla M, Agnolucci M, Avio L, Sbrana C, Turrini A, Giovannetti M. Arbuscular Mycorrhizal Fungi and Associated Microbiota as Plant Biostimulants: Research Strategies for the Selection of the Best Performing Inocula. Agronomy. 2020; 10(1):106. https://doi.org/10.3390/agronomy10010106*

Additionally, some soil bacteria have the capacity to biodegrade solvents and other hazardous organic compounds produced during the mining of minerals. Additionally, during the phytoremediation process, microorganisms are crucial in the removal or detoxification of HMs. The process of phytoremediation, which uses plants to remove contaminants from the environment, is slow despite being a sustainable and affordable solution. Therefore, it is critical to boost the effectiveness and amount of soil stabilization or elimination of harmful metal(loid)s by plants. The function of fungi in plants cultivated on metal-contaminated soils with limited nutrients, low water-holding capacity, and unfavorable environmental conditions has thus received more attention. The microbiota can aid in phytoremediation, carbon sequestration, plant productivity, and growth. Due to the small diameter of the mycelia, fungi can effectively explore soil microsites that are inaccessible to plant roots. They can also engage in competition with other microorganisms for water and metal uptake, shield the roots from direct contact with the metals, and impede metal transport by making the soil more hydrophobic. The definition of a fungus is an organism that is eukaryotic, heterotrophic, and absorptive that normally grows a mycelium or a branching, tubular body, and reproduces by sporulation. Fungi are also ideally suited for bioremediation procedures due to their capacity to build vast mycelial networks. In cases of bacterial dysfunction, the use of filamentous fungus can be a potential alternative. They are crucial for the oxidation and reduction of organic and inorganic substances, the cycling of elements, the alteration of rocks and minerals, bio-weathering, the production of mycogenic minerals, fungal-clay interactions, and metal-fungal interactions. The broad hyphal networks in the soil can also play a key role in stabilizing soil aggregates; they alter the chemical makeup of root exudates, the pH of the soil, and the bioavailability of metal(loids) in the soil.

2.8 METAL CONTAMINATION AND DETOXIFICATION

Due to their high capacity to immobilize toxic metals by either the formation of insoluble oxalate, biosorption, or chelation onto melanin-like polymers, many fungi, including Trichoderma, Aspergillus, and the arbuscular mycorrhizal fungi (AM) have shown the potential to improve phytoremediation processes in metal-contaminated soils. Vascular plants support a wide range of fungus. Plant roots are susceptible to soilborne diseases, but they are also colonized by mutualistic or non-pathogenic fungi, such as ectomycorrhizae (EM), dark septate endophytes or DSE, and AM fungi. About 150 zygomycete fungal species comprise AM fungi, while there are about 6000 species of EM fungi, the most of which are basidiomycetes with a smaller number of ascomycetes and zygomycetes (Deng and Cao, 2017). The majority of herbaceous plants and a number of woody plant families are linked with the AM fungus, whereas a smaller number of woody plant families are mostly connected with the EM fungi.

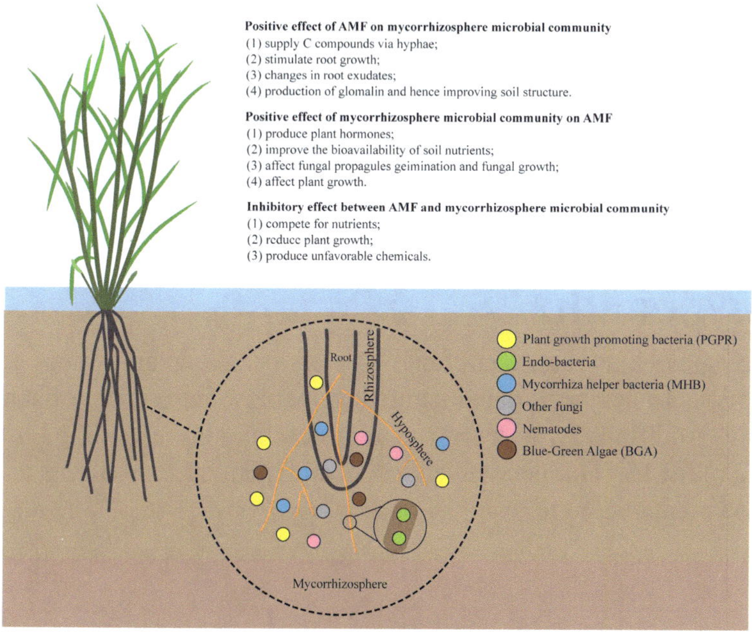

Figure 2.13. Mycorrhizal fungi and their benefits to plants.

Source: *Bao X, Zou J, Zhang B, Wu L, Yang T, Huang Q. Arbuscular Mycorrhizal Fungi and Microbes Interaction in Rice Mycorrhizosphere. Agronomy. 2022; 12(6):1277. https://doi.org/10.3390/agronomy12061277*

Mycorrhizal fungi can sequester potentially harmful metal(loid)s, aid in the transfer of nutrients from the soil to host plants and increase soil microbial activity, all of which contribute to the total biodegradation of soil pollutants (Figure 2.13). Despite being classified as non-mycorrhizal plants, certain DSEs have been discovered as fungus symbionts in Cruciferae species. The DSE are broadly categorized as conidial and sterile septate fungal endophytes that grow in plant roots and produce melanized structures such as microsclerotia and intracellular hyphae. The DSE fungi are widespread and frequently cohabit with several mycorrhizal fungi. The most prevalent soil microorganisms in agricultural and natural soils are AM fungus of the Glomeromycota. There have so far been descriptions of about 160 AM fungus taxa belonging to the order Glomales (Glomeromycota). The AM fungi were divided into 12 genera using morphological and molecular identification techniques, with the Glomus species predominating (Acaulospora, Ambispora, Archaeospora, Diversispora, Entrophospora, Kuklospora, Geosiphon, Gigaspora, Glomus, Intraspora, Paraglomus, Scutellospora) (Pinto et al, 2018). The majority of terrestrial plants have symbiotic relationships with AM fungi, which are widespread. As a result, they play a significant role in the soil microbiome and assist their hosts by increasing the soil volume. They also help their hosts acquire water more effectively and strengthen pathogen resistance. As a result, mycorrhizal plants spend less energy extending their root systems than non-mycorrhizal plants, and they are more resilient to pathogen assaults and drought stress.

2.9 MYCORRHIZA

Physiologically, mycorrhizal fungi can also make plants more resilient. Jasmonic acid has been demonstrated to increase herbivore resistance due to AMF. Additionally, mycorrhizal connections may encourage a quicker closing of the leaf stomata, which would stop plants from wilting, as well as increase plant osmolyte levels of proline.

Figure 2.14. Brassica juncea.

Source: *ht By Walther Otto Müller - List of Koehler Images, Public Domain, https://commons.wikimedia.org/w/index.php?curid=255508*

For example, AM fungal inoculation in mustard (*Brassica juncea*) (Figure 2.14) increased osmolyte content as proline and reduced overall symptoms of sodium chloride-induced drought stress (Wang et al, 2020). Mycorrhiza can help with phytoremediation by either increasing the bioavailability of metal(loid)s for plant absorption or by lowering the toxicity of metals in their host plants. However, the fungal species and ecotype have a significant impact on the effectiveness of phytoremediation. Mycorrhizal fungus must first create a mycorrhizal symbiosis in order to be able to improve the efficacy of phytoremediation. As a result, for the fungus to be used in phytoremediation applications, it must be able to live in environments that are contaminated with metal(loids). Different extracellular and intracellular defensive mechanisms are available to mycorrhizal fungi to ensure their survival in metal(loid)s-contaminated soils. Metal uptake may be inhibited by extracellular mechanisms such as chelation, cell wall binding, or biosorption. The quantity in the cytosol can be decreased by intracellular processes, such as binding to non-protein thiols and being transported into intracellular compartments. Transporter proteins and intracellular chelation, such as that provided by metallothioneins, glutathione, organic acids, amino acids, and compound-specific chaperones, are essential for intracellular processes. These metal complexes can also be transferred after being chelated. Additionally, several plasma membrane-based transporter systems that handle the uptake and release of metals can be downregulated.

Additionally, antioxidative defence systems that detoxify reactive oxygen species (ROS) as well as repairing the biomolecules damaged by metal. Their line of defence against high metal(loid) concentrations is chelation, which depends on both the metal and the species. A wide variety of various chelating chemicals secreted by fungi can be used to chelate metals. However, metal-containing chelates can cause metal solubilization when mycorrhizal fungi exude organic substances. By managing the interactions between the metals and the plant roots, mutualistic fungi can also help protect roots against the toxicity of metal(loid)s. It has been shown that mycorrhizal connections can reduce metal toxicity in higher plants. AM fungi are capable of reducing the stress caused by metal(loid) through a number of methods. Larger biomass seen in mycorrhizal plants can help to lower the concentration of metals, while AM hyphal cells can immobilize and compartmentalize metal(loid)s. Particularly, some native mycorrhizae that are still present at contaminated areas may precipitate metal oxalates in the intercellular spaces of the fungi or the host plant, which would obstruct the ability of Casparian strips to transport apoplastic molecules.

As an alternative, mycorrhizal fungi can directly defend the plant from the accumulation of phytotoxic concentrations of some pollutants by secreting specific detoxifying substances like organic acids or by encapsulating the pollutants in fungal tissues connected to the roots, thereby erecting a physical barrier against toxic metal translocation. Additionally, AM fungi can synthesize glomalin, metallothioneins, and fungal polyphosphates, all of which have significant metal(loid) binding capabilities. Glomalin can sequester metal(loid)s and is a part of the spore and hyphal cell walls of AM fungus. Glomalin is securely bound in the hyphal and spore walls of AM fungus, and very less amounts of glomalin are released by AM fungi into the soil environment. No other fungus group, except Glomeromycota, produces this glycoprotein in appreciable quantities.

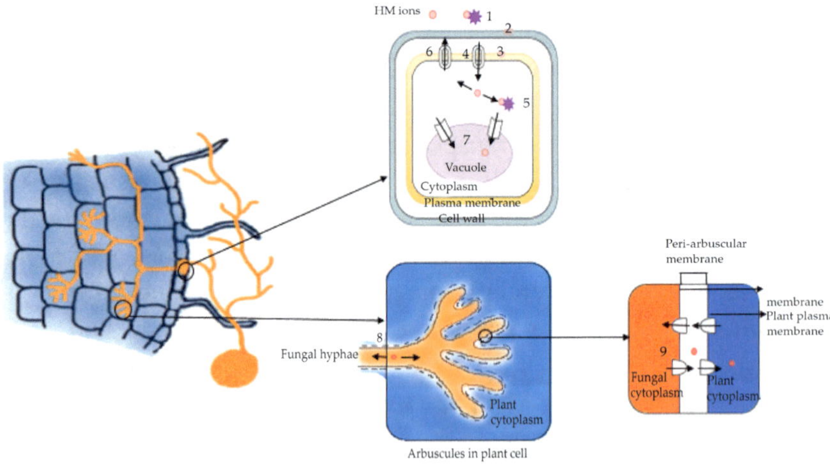

Figure 2.15. Arbuscular Mycorrhizal Fungi for heavy metals.

Source: *Dhalaria R, Kumar D, Kumar H, Nepovimova E, Kuča K, Torequl Islam M, Verma R. Arbuscular Mycorrhizal Fungi as Potential Agents in Ameliorating Heavy Metal Stress in Plants. Agronomy. 2020; 10(6):815. https://doi.org/10.3390/agronomy10060815*

The mycelium of AM fungus (Figure 2.15) plays a significant role in the sequestration of metals, even if the methods by which metal(loid)s are sequestrated by glomalin are not at all evident. This sequestration takes place both inside the hyphal cytoplasm and the mucilaginous outer hyphal wall zone. Glomalin may lessen the toxicity risk that toxic elements pose to plants and microorganisms in metal(loid)-polluted areas by stabilizing them. AM fungi frequently reduce the mobility and toxicity of metal(loids) by either raising the pH of the soil or sequestering inside extraradical mycelium (Bahraminia et al, 2016) (Bahraminia et al, 2016). The stored forms of metal(loid)s and their mobility in plant tissues are directly correlated with the phytotoxicity of metal(loid)s and the degree of plant tolerance. Therefore, one of the most significant methods by which plants detoxify metal(loid)s is through the spread of chemical forms. AM fungi may improve tolerance to Cd (*Medicago sativa L.*) (Figure 2.16) by changing the chemical forms of Cd in various plant tissues. The gene expression of host plant genes related to metal tolerance can be changed by AM fungus. This indicates that AM fungi have developed tolerance to heavy metals and that they may be key players in phytoremediation, AM fungi have also been observed in hyperaccumulators growing in metal-polluted soils. Recent research has shown that AM fungus can boost the biomass, trace element content, and metal(loid) translocation factor of hyperaccumulators.

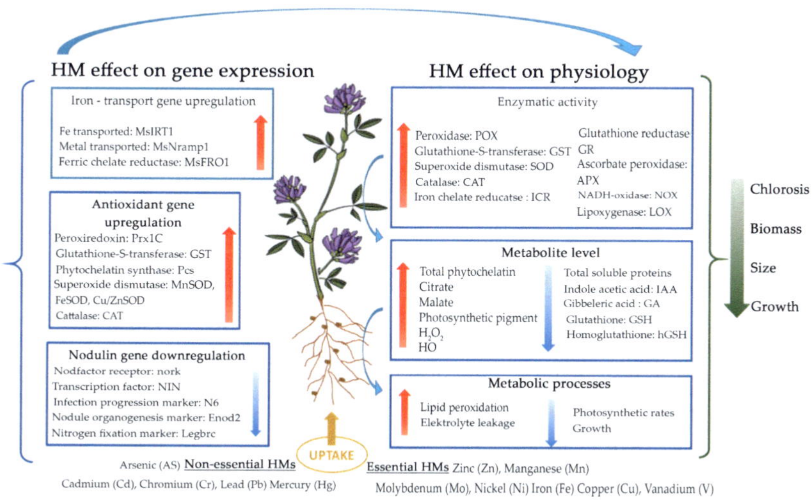

Figure 2.16. Medicago sativa and metals.

Source: *Jócsák I, Knolmajer B, Szarvas M, Rabnecz G, Pál-Fám F. Literature Review on the Effects of Heavy Metal Stress and Alleviating Possibilities through Exogenously Applied Agents in Alfalfa (Medicago sativa L.). Plants. 2022; 11(16):2161. https://doi.org/10.3390/plants11162161*

When plants were subjected to various concentrations of Cd and Pb and inoculated with a mixture of *G. mosseae* and *G. intraradices*, Cd and Pb accumulated in the roots and barely moved to the shoots. Additionally, it was discovered that AM-inoculated plants were more metal-tolerant. Thus, it was determined that inoculated *L. sinuatum* might be effective as a Cd or

Pb controlling agent for phytoremediation and that mycorrhization reduced metal toxicity in the plants. In contrast to AM fungi, endophytic fungi are far more common and include a wide variety of fungi that also have the ability to improve phytoremediation. Endophytic microorganisms, such as bacteria and fungi, are more resistant to harmful environmental changes and are more likely to interact closely with their hosts (Newman and Reynolds, 2005). To increase the effectiveness of phytoremediation, endophytic bacteria have been researched for their potential to reduce metal(loid) toxicity to plants. Therefore, metal(loid) stress that has a negative impact on plant growth can be mitigated by endophyte symbiosis. Endophytic fungi might make host plants more resistant to different metal pollution. Their increased biomass is also ideal for bioremediation, and they have suitable metal sequestration or chelation mechanisms to boost their tolerance to metal(loid)s. Additionally, the presence of metal-resistant endophytic bacteria may be advantageous for host plants because they might improve phytoextraction processes by increasing nutrient uptake and metal bioavailability via secreted metabolites in the plant rhizosphere. Endophytes are useful resources for phyto-bioremediation because of the metabolic activities they carry out, which aid in the phyto-bioremediation of pollutants and the biotransformation of resistant organic compounds through their degradative abilities.

2.10 METAL CONTAMINATED SOILS AND FUNGI

In metal contaminated soils, *Cryptococcus sp.* CBSB78 and *Rhodotorula sp.* CBSB79 endophytic yeasts have been isolated from canola roots (*B. chinensis*), and they exhibit resistance to Cd, Pb, Zn, and Cu. Endophytic fungi such as Trichoderma, Fusarium, Aspergillus, Paecilomyces, Penicillium, Paecilomyces, Cladosporium, and Lasiodiplodia are found in the Portulaca genus of several HMs hyperaccumulating plants (Deng et al, 2012). The most often isolated fungal taxa that can reduce HMs stress are *Penicillium* spp. and *Trichoderma* spp. Further, *S. nigrum* may carry fungal endophytes that might boost metal extraction and host plant growth, increasing the effectiveness of phytoremediation in the removal of Cd-contaminated soils. In fact, *S. nigrum* was used to inoculate Cd-resistant endophytic fungus, which boosted the plant's tolerance to high Cd concentrations and dramatically improved the values linked to Cd biosorption, such as the translocation factor, bioconcentration factor, and Cd tolerance index. Combining improved plant bioavailability in the rhizosphere with decreased internal bioavailability of metal(oids) should allow plants to bioaccumulate more metal(oids) while reducing phytotoxicity. A larger amount of metal(loid)s could also be absorbed by bacterial cells than by inorganic soil components such as kaolinite, vermiculite, or mica, either through an independent passive metabolism or through a metabolism-dependent active process. PGP bacteria are so named because they have the capacity to promote plant growth. These PGP bacteria may form mutualistic or symbiotic relationships with plants and can be found as free-living endophytes or rhizospheric bacteria. Among the helpful PGP bacteria that can regulate plant growth are the genera Gluconacetobacterium, Flavobacterium, Beijerinckia, Klebsiella, Erwinia, Enterobacter, Burkholderia, Pseudomonas, Serratia, and Bacillus. According to certain studies, plant growth-promoting rhizobacteria (PGPR) can increase the biomass production of plants growing in soils contaminated with metal(loids) and reduce metal-induced phytotoxicity (Venkatachalam et al., 2017).

2.11 ROOT EXUDATES AND THEIR IMPORTANCE

Root exudates can help metal-resistant bacteria on roots and in the rhizosphere, although other bacteria enter the plant as endophytes that are beneficial and can form a mutualistic relationship. Endophytic populations, like those in the rhizosphere, can be constrained by biotic and abiotic factors. In comparison to rhizospheric bacteria, endophytic bacteria may be better shielded from the effects of biotic and abiotic stressors. Endophytic bacteria can be categorized as obligate or facultative depending on how they live (Joel et al, 1996). Obligate endophytes are transmitted to other plants via vectors because they must grow and survive solely on the host plant. Facultative endophytes go through a phase in which they don't live inside host plants. Endophytic bacteria often enter plant tissues through the root zone, but they can also enter through the aerial sections of plants, such as the flowers, stems, and cotyledons. Through germinating radicles, secondary roots, stomata, or foliar damage, the bacteria enter plant tissues. Once within the plant, endophytes may either settle close to the opening or disperse throughout the tissues. These bacteria may be found inside cells, in the vascular system, or in the intercellular spaces. Despite the fact that PGPR is frequently mentioned as being important in microbially assisted phytoremediation, bacterial endophytes have many benefits over PGPR. For instance, colonizing plant roots with PGPR is frequently difficult since the inoculant must be applied at precisely the right time for the plant. Additionally, because they are in natural biocenosis, they are less competitive with indigenous bacteria.

The ability of PGPR to survive and colonize new areas is also influenced by a number of other factors, including the physiological growth of the organisms themselves and the abiotic and biotic conditions of the soil. Due to their direct contact with plants, endophytic bacteria may have a distinct competitive edge over PGPR. Furthermore, reinoculation of endophytic bacteria has no impact on the natural population of endophytes in plants. Additionally, because of their specific location within the plant, endophytic bacteria may be able to recolonize the plant surface and withstand certain biotic and abiotic stress conditions in the soil.

Metal-resistant endophytic bacteria not only shield plants from the toxicity of metal(loid)s but also increase the metal bioaccumulation in plant tissues while simultaneously stimulating plant growth. Endophytic bacteria can colonize interior tissues and create a variety of distinct lifestyles with their host, including symbiotic, mutualistic, commensalism, and trophobiotic relationships. They are found in plant tissues beneath the epidermal cell layers. They can colonize a specific host with the maximum densities in the roots and the lowest densities from stems to leaves. They are widespread in a wide variety of plant species. Most endophytes typically come from bacterial populations that live in the rhizosphere, phyllosphere, or other portions of plants, although others can spread through seeds or foliar tissues that have been injured. A closed ecosystem has developed as a result of the long-term coevolution of plants and endophytic bacteria. This ecosystem aids plants in adapting to and surviving under biotic and abiotic stress conditions, such as pathogen infection, drought, salinity, and contaminants, and improves the ecological balance of the natural system. Despite the variable and transient nature of their presence in plants, bacterial endophytes can frequently cause physiological changes that aid in the plant's growth and development. Endophytes typically have wider-ranging positive impacts than many rhizobacteria (Ma et al, 2016a).

2.12 ENDOPHYTES: THE BENEFITS

Additionally, endophytes have been linked to a range of positive impacts on plant growth, including osmotic adjustment, stomatal regulation, modification of root shape, improved mineral uptake, and changes to nitrogen buildup and metabolism. Since bacterial endophytes may aid host plants in adapting to unfavorable soil conditions and increase the effectiveness of phytoremediation by promoting plant growth, relieving metal stress, lowering metal phytotoxicity, and changing metal bioavailability in soil and metal translocation in plants, phytoremediation assisted by bacterial endophytes has been highly recommended in recent years for cleaning up metal(loid)s-polluted soils. Endophytic bacteria enhance plant growth in metal-polluted soils in two different ways: directly by producing substances that are beneficial to plant growth, such as the solubilization and bioconversion of mineral nutrients like phosphorus, nitrogen, and potassium, and indirectly by controlling plant pathogens or by inducing systemic resistance in plants against pathogens. Additionally, they alter the ability of plants to bioaccumulate metals by excreting metal-mobilizing organic acids, biosurfactants, and extracellular polymeric substances (EPS) (Tang et al, 2021). Additionally, the majority of the proteins, nucleic acids, lipids, and polysaccharides that endophytic bacteria produce outside of their cells play a substantial role in metal complexation, which lowers the bioaccessibility and bioavailability of the metals. The biosurfactants made by endophytic bacteria appeared to boost phytoremediation rates and increase the bioavailability of poorly soluble metals. Mycolic acid, glycolipids, lipopeptides, polysaccharide-protein complexes, phospholipids, fatty acids, and other chemical structures are a few examples of the diverse chemical structures that can be found in biosurfactants. The distribution of the metals is impacted by these molecules' ability to reduce surface tension, critical micelle concentration, and phase tension. The biosurfactants are made by endophytic bacteria and released to the host as root exudates. In order to modify the mobility and bioavailability of metals in the soil solution, the secreted biosurfactants first interact with complex insoluble metals on the interface of rhizosphere soil particles. They subsequently desorb metals from the soil matrix. Given that the primary function attributed to biosurfactants is their involvement in aiding metal uptake, the interactions between endophytic bacteria, metals, and biosurfactants can be interpreted from a functional perspective. Regarding indirect effects on promoting plant growth, it is known that a number of plant growth-promoting endophytic (PGPE) bacteria reduce the effects of stress on plants by preventing phytopathogen damage either through biological control of pathogens or through inducing systemic resistance (ISR) in plants against pathogens (Manoj et al., 2020).

Numerous biocontrol methods have demonstrated the ability of one or more antibiotics to contribute to the eradication of diseases. The production of antibiotics by the endophytic bacterial strains *Bacillus megaterium* BP 17 and *Curtobacterium luteum* TC 10 substantially reduced the growth of the burrowing nematode (*Radopholus similis*). The biosurfactant C-15 surfactin, which was capable of controlling the maize mycotoxic fungus

Fusarium verticillioides, was recently discovered (Chen, Juang, and Wei, 2015). Furthermore, diseases can be effectively outcompeted by endophytic bacteria for colonization niches and accessible nutrients, therefore increasing plant output. Systemically improved resistance to a wide range of infections is referred to as induced systemic resistance (ISR). The ISR to various diseases brought on by chemicals and bacteria that encourage plant development is extremely helpful in agroecosystem protection. In several plant taxa, ISR generated by endophytic bacteria has been shown to be effective against a variety of bacterial, fungal, and viral diseases. Inducing a plant defense mechanism in plants through the use of bacterial endophytes prevented diseases from colonizing the host. Reactive oxygen species are produced by such enzymatic activity, and endophytic bacteria may alter how well these organelles are protected from oxidative stress.

Endophytic bacteria might control the activity of antioxidant enzymes in plants including POS, CAT, SOD, glutathione peroxidase, and ascorbate peroxidase as well as lipid peroxidation. Recent studies with hyperaccumulator plants showed that inoculating soils, seeds, and seedlings with metal-resistant endophytic bacteria accelerated the phytoremediation process in naturally and/or artificially metal(loid)s-contaminated soils by improving nutrient uptake, cell elongation, metal bioaccumulation or stabilization, and alleviating metal stress in plants. Similar to this, it is well known that the colonization and spread of plant growth-promoting endophytic bacteria contribute to the improvement of soil fertility and the stimulation of host plant development. They do this by supplying a variety of growth regulators and essential nutrients or by synthesizing 1-aminocyclopropane-1-carboxylic acid (ACC) deaminase. Metal resistance, detoxification, accumulation, biotransformation, sequestration, and other multiple plant growth-promoting traits of such endophytic bacteria can reduce phytotoxicity and alter the bioavailability of metal(loid)s in contaminated soils, making them an ideal choice for studies on microbial-assisted phytoremediation.

Numerous investigations have shown that endophytic bacteria isolated from hyperaccumulator plants showed stronger metal tolerance than those from non-hyperaccumulator plants, despite the fact that metal(loid)s negatively influences endophytic bacterial diversity in plants. The impact of various metal(loid)s on endophyte diversity, biomass, and activity has been studied. But it is widely known that bacteria separated from polluted settings can withstand higher metal concentrations than bacteria isolated from clean environments. The death of sensitive species and the ensuing competition and adaptation of remaining bacteria after the addition of metals increases metal tolerance in bacterial communities. Horizontal gene transfer (HGT) mediated by plasmids harboring metal resistance genes may be the cause of bacterial metal(loid) resistance. The mercury resistance (Me) operon, which is primarily found in gram-negative bacteria, is organized with the genes describing the numerous actions required for the biodegradation of organomercurials and reduction of Hg^{2+}. Similar to rhizobacteria, PGPE promotes host plant growth directly in a number of ways. The mechanisms are nitrogen fixation, mineral solubilization, phytohormone production, specialized enzyme production, and siderophore production. Diazotrophs (Figure 2.17) are bacteria that can fix nitrogen, or change stable atmospheric nitrogen gas into a form that is useful to living things.

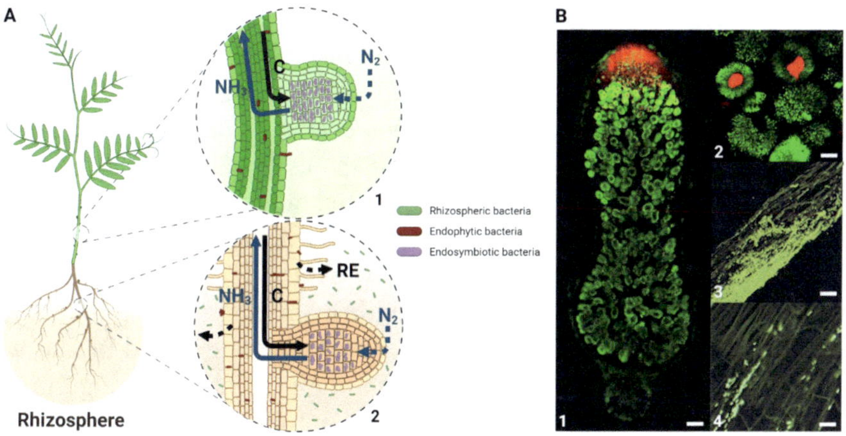

Figure 2.17. Diazotrophs and plants.

Source: *Mandon K, Nazaret F, Farajzadeh D, Alloing G, Frendo P. Redox Regulation in Diazotrophic Bacteria in Interaction with Plants. Antioxidants. 2021; 10(6):880. https://doi.org/10.3390/antiox10060880*

Diazotrophs use the nitrogenase enzyme to convert dinitrogen to ammonia. Compared to other rhizospheric microbes, endophytic bacteria have a significant impact on boosting plant health and growth and enable plants to thrive in nitrogen-poor soil settings. One of the most important nutrients for plants is phosphorus (P), which has a significant impact on a number of vital metabolic processes, including cell division and growth, energy transport, signal transduction, macromolecular biosynthesis, photosynthesis, root development, and respiration of plants. In agricultural soils, P is therefore frequently viewed as a limiting nutrient. Some metal-resistant endophytic bacteria have been found to mineralize organic phosphorus in soil by secreting extracellular acid phosphatase or to solubilize precipitated phosphates in soil by acidification, ion exchange, and release of chelating agents under metal(loid)s stress conditions, thereby improving P availability to plants. Endophytic bacteria are well known for assimilating soluble P and preventing its subsequent sorption. Therefore, even in phosphate-limiting situations, these endophytic bacteria may behave as a sink for phosphorus by fast microbial phosphorus mobilization. As soon as the phosphorus is released from the bacterial cells, these endophytes turn into a source of phosphorus for the plant hosts.

The conversion of insoluble forms of P to bioavailable forms is facilitated by the presence of phosphate solubilizing bacteria in the rhizosphere as well as the production of organic acids and phosphatases. Several endophytes have been identified as effective P solubilizers, including *Pseudomonas citronellolis, Pseudomonas oryzihabitans, Enterobacter agglomerans, Pantoea sp., Enterobacteriaceae, Burkholderia, Ralstonia pickettii, Erwinia sp.,* and *Klebsiella pneumoniae*. One of the essential elements for life is iron (Fe), and almost all organisms except some lactobacilli require iron-containing proteins for a variety of physiological functions and enzymatic reactions. However, iron resides mostly as Fe(III) in the aerobic soil environment and frequently forms insoluble hydroxides and oxyhydroxides, rendering it essentially inaccessible to microbes. Bacteria had to find ways to solubilize the metal before absorption in order to obtain enough iron. In addition to rhizosphere bacteria, numerous plant species have yielded metal-resistant endophytic bacteria that also produce siderophores. A wide variety of bacteria, including nitrogen-fixing free-living organisms and symbiotic partners as well as plant and animal diseases, produce siderophores. The majority of siderophore-producing bacteria, however, are found in the rhizosphere of plants that promote growth. These bacteria grow and produce siderophores under adverse environmental conditions, such as nutrient scarcity or high metal(loid) concentrations, and may be particularly helpful for phytoremediation. Interveinal chlorosis in younger leaves, a symptom of iron deficiency brought on by an excess of metal(loid)s, has been observed in a variety of plant species. Their low iron level typically prevents the development of chloroplasts and the manufacture of chlorophyll, which hinders plant growth. Under these conditions, siderophore-producing bacteria such as rhizospheric and endophytic bacteria may provide a biological rescue system that can scavenge Fe^{3+} and make it accessible to plants lacking in iron.

2.13. CHELATING SUBSTANCES AND IRON BIOAVAILABILITY

Microbial synthesis of chelating substances such as siderophores that can solubilize Fe in iron deficient situations may alter the iron bioavailability to plant roots. Low-molecular-weight organic molecules called siderophores have an affinity to Fe^{3+} ions and can also bind other metal ions that plants can absorb. Two fundamental methods are used to acquire iron (Gavrilescu, 2022). After root-mediated biodegradation of the chelate, the first method involves the uptake of Fe in Fe-siderophore complexes. Since microbial siderophores typically have a higher affinity for iron than phytosiderophores, plants growing in metal-contaminated soils are able to accumulate high amounts of iron with the aid of siderophore-producing bacteria. The second strategy involves the solubilization of unavailable forms of iron by the release of phytosiderophores. Therefore, it is believed that under situations of metal stress, bacterial siderophores will be the primary providers of phytoavailable Fe for plants. Therefore, when plants are cultivated in metal-polluted soils, inoculating them with bacteria that can produce siderophores may help to prevent them from going chlorotic.

Therefore, in situations when there is a shortage of iron, siderophores function as substances that solubilize iron from minerals or organic compounds. In addition to iron, siderophores can also produce stable complexes with other potentially hazardous metals like Al, Cd, Cu, Ga, In, Pb, and Zn as well as radionuclides like U and Np. There are currently over 500 distinct siderophores known. The functional groups that coordinate the iron atom are not as diverse while having vastly different overall structures. Siderophores can be categorized as belonging to the hydroxycarboxylate, catecholate, or hydroxamate-type siderophores depending on the hydroxycarboxylic acid, catechol, or hydroxamic acid moieties that are present in their metal-binding sites. In most cases, siderophores combine 1:1 with Fe^{3+} to form complexes, which are then absorbed by bacterial cell membranes, where Fe^{3+} is converted to Fe^{2+} and discharged from the siderophore into the cell. The siderophore is occasionally destroyed during this reduction, but occasionally it is recycled. Both gram-negative and gram-positive bacteria have this mechanism of iron uptake from siderophores. External pH is crucial for the synthesis of microbial siderophores as well as for their chemical stability in the environment, even though siderophores are largely produced in response to iron deficiency. In general, hydroxamate siderophores are common in acidic soils and show a preference for low pH values for hydroxamate siderophores synthesis by microbes. Contrarily, the growth of catecholate siderophores is aided by neutral to alkaline soils. Heavy metal concentrations in the environment can affect siderophore biosynthesis and microbial growth in addition to pH.

It has been demonstrated that bacterial strains isolated from polluted settings are more metal-tolerant than those obtained from unpolluted locations. These bacteria have developed a number of methods to survive metal stress. The stimulating effect of heavy metals on siderophore synthesis has at least two potential explanations. First, the siderophore production processes or their control may directly include the heavy metal. Alternatively, complex formation with metal ions may lower the concentration of free siderophores in the medium.

This process prevents siderophores from complexing with iron, lowering the concentration of soluble iron as a result. More siderophores would then be produced since iron deficiency enhances siderophore production. Several plant species, including *Cucurbita pepo, B. juncea, Helianthus annuus, Medicago sativa, and Vigna unguiculata*, can acquire iron through Fe-siderophore complexes (Grobelak and Hiller, 2017). It was also demonstrated that this acquisition was the cause of their increased growth in metal-contaminated soils. Endophytes are thought to produce phytohormones, which are thought to be crucial for plant-bacterial interactions and plant growth in metal-contaminated soils.

Endophytic colonization frequently results in increases in nutrient absorption and plant biomass. These would imply that endophytic bacteria's ability to reduce metal(loid)s stress is the consequence of a combination of nutritional and metabolic advantages.

In general, phytohormones include indole-3-acetic acid (IAA), cytokinins, gibberellins, abscisic acid, and ethylene, which, depending on the substance quantity, may either inhibit or promote growth. IAA, an auxin, participates in a variety of physiological activities in plants, including regulating plant development, triggering plant defence mechanisms, and acting as a cell-cell signaling molecule.

IAA creates a gradient of concentration in various plant sections. Tissue differentiation may happen as a result, depending on its concentration, whether it inhibits or stimulates growth. The amount of IAA emitted by endophytes is now believed to be crucial in regulating the interaction between plants and endophytes as well as plant development in metal(loid)-contaminated soils.

2.14 PHYTOHORMONES

It was discovered that strains of Pseudomonas, Enterobacter, Staphylococcus, Azotobacter, and Azospirillum generate auxins and cytokinins. Together with gibberellins, these chemicals have the potential to impact plant development and growth. For example, Azospirillum-inoculated roots demonstrated increased root cell membrane activity, free IAA and indole-3-butyric acid levels, as well as specific activities of the tricarboxylic cycle and glycolysis pathway. Root elongation may change qualitatively in accordance with the IAA level. Primary root elongation is encouraged by low levels of the phytohormone produced by bacteria, whereas primary root development is inhibited by high levels of IAA but is stimulated by lateral and adventitious root formation (Djedidi et al., 2014). Endophytes can therefore promote plant development by changing the hormonal balance of the plant. For instance, *P. fluorescens* is typically thought of as a rhizosphere bacterium that mostly colonizes the elongation and root hair zones of roots. The roots and stems of several hyperaccumulators contain *P. fluorescens*, which can also be endophytic. The use of ACC as the exclusive source of N is another way that the endophytic bacteria may have an impact on the growth of the host plant. Production of phytohormones has been linked to endophytic bacteria that are typically advantageous to plants in situ, such as Azotobacter, Azospirillum, Pseudomonas, Enterobacter, Staphylococcus, etc. Both bacterial IAA and ACC deaminase may play a role in the accelerated proliferation of the Cd hyperaccumulator *S. nigrum L.* caused by bacterial endophytes under cadmium stress. As a common plant hormone, ethylene is crucial for root initiation and nodulation, cell elongation, leaf senescence, abscission, fruit ripening, as well as auxin transport in plants' responses to abiotic and biotic stressors. The following enzymes are involved in the production of ethylene in higher plants: S-adenosyl-L-methionine (SAM) synthetase catalyzes the conversion of methionine and adenosine triphosphate (ATP) to SAM; ACC synthase mediates the hydrolysis of SAM to ACC and 5′-methylthioadenosine (MTA). Plants under metal(loid) stress produces more ethylene, which inhibits root elongation, lateral root growth, and root hair development. In these conditions, specific endophytic bacteria may reduce the effects of stress on plants through the enzymatic hydrolysis of ACC (Mata et al, 2022).

From seed germination to shoot growth and leaf abscission, ethylene is involved in almost every phase of plant growth and development. Therefore, the ability of endophytes to produce ACC deaminase is probably crucial and effective for controlling their plant hosts. The ACC deaminase enzyme can hydrolyze ACC into ketobutyrate and ammonia, which bacteria can easily use as a source of nitrogen for metabolism. ACC deaminase-producing endophytic bacteria are typically found in the apoplast of plant roots. These microorganisms serve as a sink for ACC. The bacteria promote the growth of plant roots and shoots while decreasing the negative effects of ethylene production by lowering ethylene levels. The proportion of endophytes that could use ACC as the only N source was higher than that of rhizosphere bacteria. *Methylobacterium oryzae* and *Burkholderia sp.*, which were isolated from rice tissue, were used in experiments to demonstrate their capacity to lower ethylene levels and shield tomato (*Lycopersicon esculentum*) from the toxicity of high Ni and Cd concentrations. The ability to combat metal phytotoxicity is crucial for the effectiveness of phytoremediation. In order to reduce metal toxicity or increase plant metal tolerance, many bacterial-mediated processes are implicated in the endophyte-host coevolution process. The endophytic bacterial strain *Bacillus sp.* MN has developed a more well-defined metal-resistant mechanism, such as active export via a P-type ATPase efflux pump, which can move metal ions across biological membranes against the concentration gradient using the energy released during ATP hydrolysis. In plant-endophyte mutualistic systems, endophytic bacteria with unique metal bioaccumulation capacities can be employed to speed up the detoxification of metal(loid)s and boost the effectiveness of phytoremediation. It is well known that the bioavailability of metal(loid)s, which can be influenced by a number of variables including soil particle size, nutrients, soil pH, redox potential, organic matter content, and the presence of other ions, is what determines how readily plants can absorb metal(loid) s from soils. Numerous studies have suggested that endophytic bacteria with a metal resistance/ sequestration pathway, such as ncc-nre, can reduce phytotoxicity and increase the bioavailability of metal(loid)s through the release of metal chelating agents, such as siderophores, biosurfactants, and organic acids, soil acidification, redox activity, and phosphate solubilization (Montes-Bayón et al., 2002).

Many researchers are currently interested in the interactions between endophytes and hyperaccumulator plants because of the biotechnological potential for bioremediation and to examine the composition of bacterial communities living in naturally polluted environments. A particular environment for bacterial endophytes adapted to thrive in high metal concentrations can therefore be provided by hyperaccumulating plants, which often accumulate enormous amounts of metal(loid)s. *Alyssum bertolonii, Thlaspi caerulescens, Thlaspi goesingense,* and *Nicotiana tabacum* are a few examples of hyperaccumulating plants from which metal-resistant endophytic bacteria have been discovered. Isolates from the shoot and root showed various tolerances, indicating that various microbial communities occur in various plant compartments. Similar mechanisms to those outlined for PGPR appear to underlie endophytes' positive impacts on their hyperaccumulators. Also found to be common soil bacteria are numerous endophytic bacterial taxa from hyperaccumulators. Two steps are involved in bacterial metal biosorption: Living and dead/inactive cells passively biosorb metals through a variety of metabolism-independent mechanisms that fundamentally take place in the cell wall. In this process, interactions between metals and functional groups on the cell surface, such as hydroxyl, carbonyl, carboxyl, sulfhydryl, thioether, sulfonate, amine, amide, and phosphonate, sorb quickly to the cell surface. Ion exchange, complexation, coordination, sorption, chelation, electrostatic interaction, and microprecipitation are a few examples of several metal-binding mechanisms that may work together or separately. Active biosorption, also known as bioaccumulation, is the process by which metals are taken up by cells. This only happens in living cells due to a delayed metal transfer into bacterial cells that is dependent on active metabolism. Depending on the organism and the metal in question, the metals may be bound, precipitated, aggregated, sequestered inside particular intracellular organelles, or translocated to specific structures once they are inside living cells.

Endophytic bacteria have developed various mechanisms, including the efflux of metal ions from cells and subsequent extracellular precipitation, intracellular accumulation, and biotransformation of toxic metal ions into less or non-toxic forms to avoid metal stress. Metals that bind to extracellular substances can become immobile and be prevented from entering cells. For instance, different metal(loid)s can interact with the anionic functional groups such as the sulfhydryl, carboxyl, hydroxyl, sulfonate, amine, and amide groups found on the surface of cells. Similar to proteins, polysaccharides, and humic substances, microbial extracellular polymers can effectively bind heavy metals. Thus, these compounds either detoxify metals by generating simple complexes or by creating a strong barrier around the cell. Additionally, a lot of bacteria mediate processes or produce metabolites that cause metals to precipitate and crystallize. It is known that many bacteria have transporters that facilitate the efflux of metal(loid)s from cells via ATPase pumps or chemiosmotic ion/proton pumps in the event of toxicity or high concentration. Because these transporters have a high substrate affinity, they are able to reduce the amount of metal in the cytosol. For defence against metal(loid)s, certain bacteria have evolved a cytosolic sequestration mechanism. Metal ions may compartmentalize or change into less harmful forms once they have entered the cell. The corresponding microorganisms may be able to accumulate higher intracellular concentrations as a result of this process, which can function as an efficient detoxifying mechanism. Examples include the production of proteins that bind to metals, like metallothioneins. These proteins have low molecular weights and strong affinities for metals like cadmium, copper, silver, lead, mercury, etc. Metal presence triggers their production, and their main job is to detoxify metals. Additionally, some bacteria use methylation as a detoxification or metal resistance mechanism. Only some metals are methylated, hence this process is thought to be metal-dependent. Methyl groups are transferred to metals and metalloids in this process. The toxicity of selenium metalloid is reduced upon methylated.

2.15 MICROBES AND BIOREMEDIATION

Total culturable numbers, a quantitative measure of soil microbial populations, revealed a sharp decline in the diversity of microbial groups in contaminated soil samples as compared to uncontaminated samples according to a study. But compared to other culturable heterotrophic bacteria and asymbiotic nitrogen fixers, actinobacteria showed less sensitivity. Due to the complexity of their physiological and metabolic processes, actinobacteria have tools for the bioremediation of a variety of pollutants, including oil, rubber, plastics, pesticides, and heavy metals, among others. The true global issue is co-contamination. Since metal(loid)s would limit biodegradation, environments contaminated with inorganic and organic substances are thought to be challenging to bioremediate. However, recent studies have shown that certain actinobacteria strains may remove both HMs and pesticides at the same time. Although the bioremediation abilities of Streptomyces, Rhodococcus, and Amycolatopsis are among the most investigated taxa, integrated omic methods have never been able to support these abilities (Mawang et al, 2021). The usefulness of using omic technologies depends on the possibility of enhancing actinobacterial bioremediation processes using engineering techniques. It may be important to design biotools that are resistant to both organic and inorganic harmful substances based on the existence of co-contaminated settings. The process of contaminant uptake by plant roots from soil, water, wastewater, or sediments and their translocation and accumulation in biomass, such as shoots or any other harvestable plant parts, is known as phytoextraction. This process is also referred to as phytoaccumulation, phytoabsorption, or phytosequestration. In contaminated locations, metal(loid)-accumulating plants are grown, and the metal-rich biomass is harvested, removing some of the contamination. The primary and most effective phytoremediation method for removing heavy metals and metalloids from contaminated soils is phytoextraction. It is also the most popular and promising method of reclaiming soil for use in industry.

The volume and transportation costs are reduced by compaction and composting, but the amount of dissolved metal and organic compounds that leach increases. The commercial use of contaminated biomass as a source of energy is one of the most cost-effective methods for handling it. Important trends in the generation of thermal and electric energies include gasification and combustion. With a reduction in biomass to 3-6 percent and the proper location of the ash, direct combustion should be done safely. The conversion of plant biomass into biogas through thermal processing or thermochemical gasification is a sophisticated, high-tech process that takes place at 800-1300 °C. The procedure of extracting the ash is known as phytomining, and the produced ash can be used as biore. Thermochemical processing of plant biomass or pyrolysis, which involves the thermal degradation of organic molecules in an oxic environment at relatively low temperatures, is another efficient technique. This procedure enables the collection of gases, coke, and fluid oils like biofuel, pyrofuel, and resins. Coke is valuable as a fuel in industry and retains heavy metals. Metal speciation controls how well plants can collect metals from the environment through their roots. It is well known that plants' ability to absorb metals may be slowed down by the low bioavailability of soil metal(loid)s. The bioavailability of the target metal(loid)s and the capacity of different plants to bioaccumulate metals within their biomass are the two main limits of the majority of metal phytoextraction techniques. The percentage of metal in the soil that can interact with a biological target is known as metal bioavailability. Elements can be found in the soil solution as free ions, pairs of ions, ions complexed with organic anions, ions complexed with organic macromolecules, and ions complexed with inorganic colloids. The metals complexed by organic matter sorbed onto or occluded within oxides, and clay minerals coprecipitated with secondary pedogenic minerals like Al, Fe, Mn oxides, carbonates and phosphates, sulfides, or as part of the crystal lattices of primary minerals, are the most significant metal pools in the solid phase (Kidd et al., 2009).

2.16 PHYTOEXTRACTION

The process of phytoextraction may be improved by the addition of soil amendments that can boost trace element bioavailability in the soil. Current phytoextraction procedures either use hyperaccumulators or fast-growing, high biomass plants. Chelating compounds, including synthetic aminopolycarboxylic acids, can be applied effectively. The phytoextraction of Cu, Cd, Pb, Zn, and Ni can be improved by these substances. Chelating chemicals have a few drawbacks, including negative effects on soil microbes, potential groundwater contamination, and breakdown of synthetic organic acids (Entry et al., 1996). Metal removal may carry on for a very long time after the use of chelate-forming chemicals. There is mounting evidence that the effectiveness of metal phytoextraction can be determined by plant-microbe interactions and dynamics. It is well known that one way to increase the effectiveness of phytoextraction is by inoculating the plant's rhizosphere with microorganisms. In general, microorganisms can enhance phytoextraction by boosting plant biomass and the bioavailability of metal(loid)s to the plant. The inoculation of advantageous bacteria into the plant rhizosphere is a widely used technique to enhance phytoextraction. The effectiveness of trace element phytoextraction can also be increased by increased plant biomass production. Increased trace element uptake can be attributed to microbial modifications of the absorptive properties of the roots, including increased root length, surface area, and the number of root hairs, or by increasing the bioavailability of trace elements in the rhizosphere and their subsequent translocation to shoots via positive effects on plant growth, trace element complexation, and reduction of phytotoxicity.

AM fungi are known to help their host plants by changing the biological regulation of their phytoextraction efficiency, which is one of the several microbe-assisted techniques. AM fungi are frequently added to the soil for land reclamation and are considered to be one of the most significant soil microbial groups that affect metal uptake by plants and metal immobilization in soils. AM fungus inoculation could enhance host plant biomass and metal concentrations, leading to an increase in the uptake of metal(loid)s. Different host plants, fungus species, and environmental factors can alter plant response mechanisms. Plant growth, element absorption, and enzyme activity are all considerably influenced by functional diversity. In certain instances, AM fungus instead of the host plants' above-ground tissues contributed to a large storage of metal(loid)s at the root level (phytostabilization). However, in other instances, AM fungi supported successful phytoextraction by enhancing absorption and translocation to shoots. The choice of suitable AM species is crucial when treating contaminated soil using mycorrhizal plants. The most effective species are those that can survive in environments that are heavily contaminated by metal(loid)s; as a result, they may work more effectively in comparison to other AM species. These communities are metabolically and taxonomically diverse, with microorganisms that are already adapted to the environment in which they live. These microorganisms are also capable of metabolic activities that can change the bioavailability of metal(loids) and encourage plant growth.

CHAPTER 3

Wetland Contamination and Phytoremediation

CONTENTS

3.1 INTRODUCTION

Water is a crucial element of the environment and, as such, is understood to be essential to many biological processes and the persistence of life on earth. The key characteristics of water are found in its usual chemical composition. At room temperature, it is an odorless, tasteless, and colorless liquid. Water exhibits extremely complex chemical and physical properties that are still not fully understood. It has a boiling point of 100°C (212°F). Its melting and boiling points are substantially greater than predicted compared to comparable substances like hydrogen sulfide and ammonia (Rai, 2008). Another peculiar characteristic of water is that it is less dense in its solid state (ice) than in its liquid state. The majority of the human body is made up of water, and removing it results in instant death. Because it is required at some point in every stage of life, water is crucial for the completion of the life cycle in both animals and plants. For all living things to survive, water is necessary. In oceans, complex aqueous solutions in which a variety of elements are dissolved, are also thought to have been the first places where life first appeared. Aqueous solutions, like blood and digestive juices, are used by living things as a medium for biological activities. The distinctive characteristics of density, high specific heat, and liquid-solid state enable the establishment of a stratified environment that significantly regulates the chemical and biotic characteristics of lakes. Water offers a stable environment where variations in water availability and temperature are lessened in comparison to the challenges faced by aerial biota.

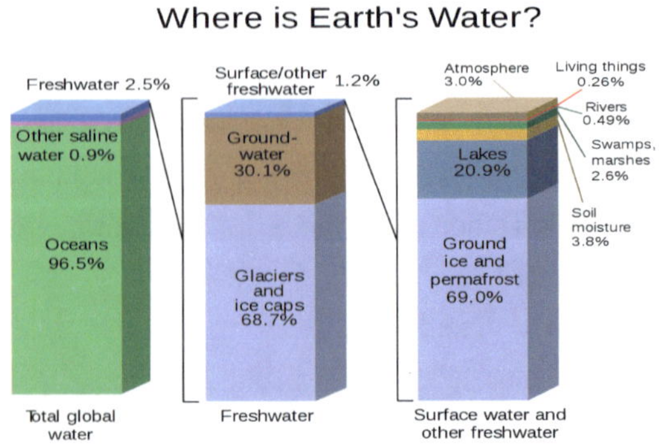

Figure 3.1. Freshwater and its availability statistics.

Source: *https://upload.wikimedia.org/wikipedia/commons/thumb/5/58/Earth%27s_water_distribution.svg/1200px-Earth%27s_water_distribution.svg.png*

3.2 FRESHWATER

Freshwater (Figure 3.1) is found in rivers, lakes, ponds, groundwater, and soil moisture, making up just 0.9 percent of the world's total water reserves, compared to the 97% (ocean), 2.1 percent (ice caps), and 2% (glaciers). Less than 1% of the available water can support human society, therefore its effective conservation and management are requirements that are being negatively impacted by a variety of developing contaminants.

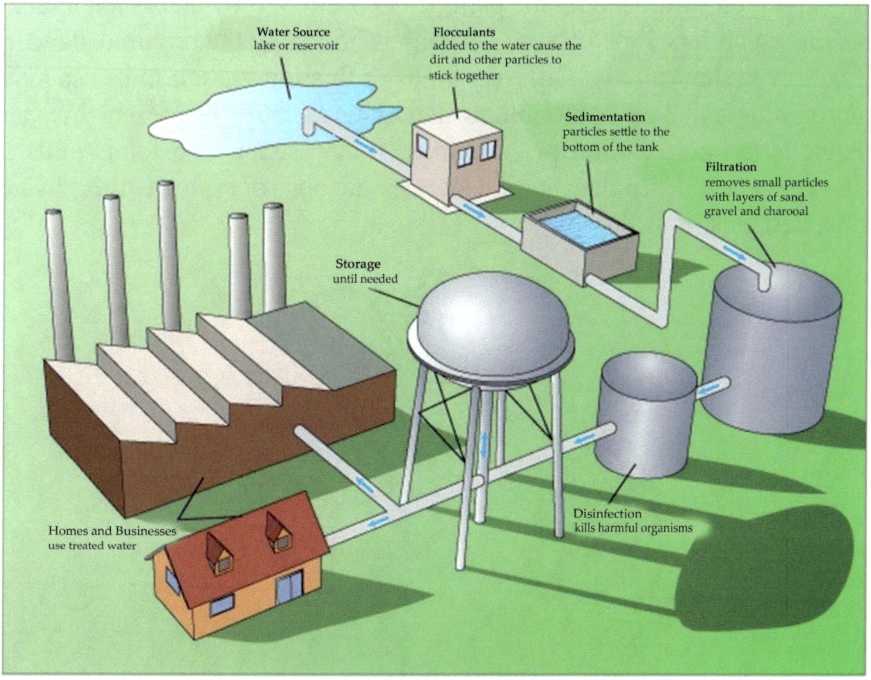

Figure 3.2. Purification and recycling of water

Source: *https://commons.wikimedia.org/wiki/File:Illustration_of_a_typical_drinking_water_treatment_process.png*

Purification and recycling of water (Figure 3.2) become more crucial as the world's population and need for freshwater rise. It is interesting to note that water purity standards for industrial use frequently exceed those for human use. Oceanic water has a greater presence but serves less purpose directly (Kösesakal et al, 2016). Freshwater resources have been crucial to the growth and development of civilizations. Humans use freshwater resources for a variety of purposes, including drinking, cleaning, transportation, and irrigation. Therefore, effective conservation and management are necessary in order to maintain this small amount of usable water for many uses that are essential to human survival. The most significant freshwater resources on the planet are lakes and surface water reservoirs, which offer countless advantages. They are vital resources for industry because they provide water for irrigation, residential usage, and renewable energy in the form of hydropower.

It is necessary to examine water using many physicochemical parameters. The only factor influencing the parameters used to test water is the intended use for which it will be utilized, as well as how high the required standards of quality and purity are. Different kinds of floating, dissolving, suspended, microbiological, and bacteriological pollutants may be present in water. Temperature, color, odor, pH, turbidity, transparency, and total solids are a few physical tests that should be carried out to evaluate the physical characteristics.

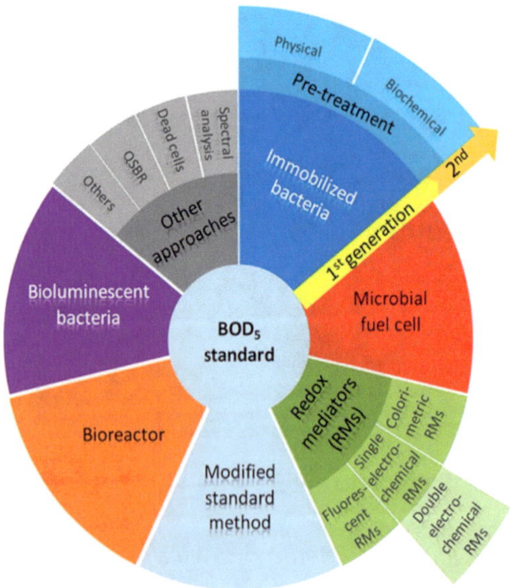

Figure 3.3. Biological oxygen demand.

Source: *https://ars.els-cdn.com/content/image/1-s2.0-S0043135413008920-fx1.jpg*

Chemical tests should be carried out to evaluate the biological oxygen demand (BOD), (Figure 3.3) dissolved oxygen (DO), acidity, alkalinity, chloride content, hardness, nitrate content, and phosphate content. Water should be analyzed for trace metals, heavy metal concentrations, and organic residues in order to obtain more potable or pure water. All these criteria are only properly monitored in wealthy countries. Water contains very low levels of organic pesticides and heavy metal contaminants, necessitating the use of very advanced testing equipment and skilled labor. Numerous physical, chemical, and biological processes in natural aquatic systems are impacted by temperature. The biologically significant effect of temperature is the reduction in oxygen solubility with rising temperature. As a result, biological species like fish and aquatic plants may require more oxygen as a result of a rise in temperature. Additionally, as a result of being exposed to higher temperatures, chemical compounds usually become more soluble. Although climatic factors play a major role in regulating temperature, human activities can also have an impact. The location and species composition of aquatic communities might suffer from thermal or chemical pollution.

The ability of inland water to reduce seasonal and diurnal temperature fluctuations both inside aquatic habitats and, to a lesser extent, beyond them, is one of the main effects of this capability to absorb heat on the biosphere. The majority of the heat added to inland waters comes from the sun. Reflection and backscattering cause a large portion of solar light that reaches the water column surface to be lost.

The remaining components reach the water column, where they are quickly absorbed and transformed into either heat through physical processes or chemical energy through photosynthesis in living things. Being a solvent is one of the most important characteristics.

Water is regarded as a universal solvent. It has an unparalleled ability to dissolve in solution in an incredibly wide variety of substances, including electrolytes such as salt, which dissociates into ions in aqueous solution, colloids, and nonelectrolytes like glucose, which retain their molecular structure and do not dissociate into ions. In inland waters, a wide range of dissolved material combinations can exist. The cations sodium, potassium, calcium, and magnesium as well as the anions chloride, sulfate, and bicarbonate or carbonate are the main inorganic solutes. Inland waters are typically thought of as fresh when the total concentration of these ions, that is, the salinity, or salt content is less than 3 g/L (Athapattu et al, 2016). All inland waters also contain smaller amounts of other ions, including the essential plant nutrients like phosphate and nitrate, as well as oxygen, carbon dioxide, and nitrogen, whose solubilities are inversely correlated with temperature, altitude, and hydrogen ion concentration (pH), and which are important for biological processes.

3.3 PHYSICOCHEMICAL PROCESSES AND INLAND WATER BODIES

Every inland body of water is affected by physicochemical processes, which result in different interactions between the biotic and abiotic elements of the ecosystem. All critical components move through inland waterways over multiple, intricate, clearly defined, and frequently interconnected biogeochemical pathways. The most notable physicochemical characteristics of inland waterways include qualities that are constantly changing and have an impact on the hydrological behavior and bioavailability of developing pollutants. The inland waters are where significant biological diversity, intricate biogeochemical processes, and energy activities take place. Inland waterways may only make up a minor portion of the biosphere geographically, but when seen from an ecological perspective, they are recognised to play a significant role in the biosphere's diversity, structure, and function. Differences in density play a significant role in controlling the overall physical and chemical dynamics of lakes as well as the metabolism that follows. A roughly linear relationship exists between density and dissolved salt content.

Figure 3.4. Deforestation: Fure River Forest Reserve Globally Significant Biodiversity Areas (GSBAs) turned into surface mining

Source: *Frank Ankomah, Boateng Kyereh, Michael Ansong, Winston Asante, "Forest Management Regimes and Drivers of Forest Cover Loss in Forest Reserves in the High Forest Zone of Ghana", International Journal of Forestry Research, vol. 2020, Article ID 8865936, 14 pages, 2020. https://doi.org/10.1155/2020/8865936*

Because of the numerous man-made activities that continue to pollute water, including rapid population growth, alarming rates of industrialization and deforestation, (Figure 3.4) urbanization, rising living standards, and a wide range of other human activities, clean water is becoming more and more valuable. Water pollution is caused by an unwanted foreign substance present in excessive amounts in the water. Given the importance of water to people, animals, and aquatic life, water pollution is one of the most severe issues faced by people today. It is the chemical that has the power to dissolve or suspend a wide range of harmful substances, primarily from the heavy flux of sewage, industrial effluents, and household and agricultural waste. Studying water pollution is particularly interesting because of this. A cost-effective and eco-technological solution to the global problem of water contamination of wetlands caused by various new pollutants is still needed. Endocrine disruptors, medicines, and personal care products (PCPs) are examples of developing micropollutants, which are pollutants with no established regulatory limitations (Ekperusi, Sikoki, and Nwachukwu, 2019). The environment and water have been disturbed during the past few decades by a variety of chemicals and pollutants, including pathogenic microorganisms, insecticides, volatile organic carbons, pesticides, personal care products (PCPs), and heavy metals. A global study has begun on the search for alternative, ecologically friendly phytotechnologies in light of a number of drawbacks and side effects of chemical and other traditional technologies. Plants can be a better tool for illuminating the mechanisms of phytoremediation, which removes developing toxins by removing them from the soil, air, and water interfaces. This is why the use of wetland plants or macrophytes in the phytoremediation of emerging pollutants is a passive, solar-driven, eco-friendly, visually beautiful, and cost-effective method of environmental management. As a result, phytoremediation is now a widely accepted method for removing developing toxins from the environment.

3.4 WETLANDS-THEIR SIGNIFICANCE

Wetlands not only provide a home for birds but also a nursery for many different types of fish, shellfish, and other organisms. Coastal wetlands have a high species diversity, particularly since they are an ecotone between freshwater and terrestrial environments or freshwater and marine habitats. Wetlands also serve crucial ecological purposes. They control the water regime, serve as organic filters, and exhibit remarkable nutrient dynamics. Wetlands serve as an ecosystem that can recycle nutrients, release surplus nitrogen, deactivate phosphates, treat wastewater, and absorb pollutants, chemicals, organic pesticides, and heavy metals through plant growth. Mangrove wetlands serve as a barrier against destructive storms in coastal areas. Wetlands, therefore, aid in preventing floods, replenishing aquifers, and lowering surface runoff and the ensuing erosion. Wetlands enhance the water's quality and maintain a comfortable climate. Urban wetlands are a natural wastewater collection system and an efficient way to use the nutrients in waste for agriculture and fisheries. Chemical activities are stimulated by the shallow depth and full exposure to sunlight in wetlands, whereas bacterial processes are stimulated by larger input due to higher productivity.

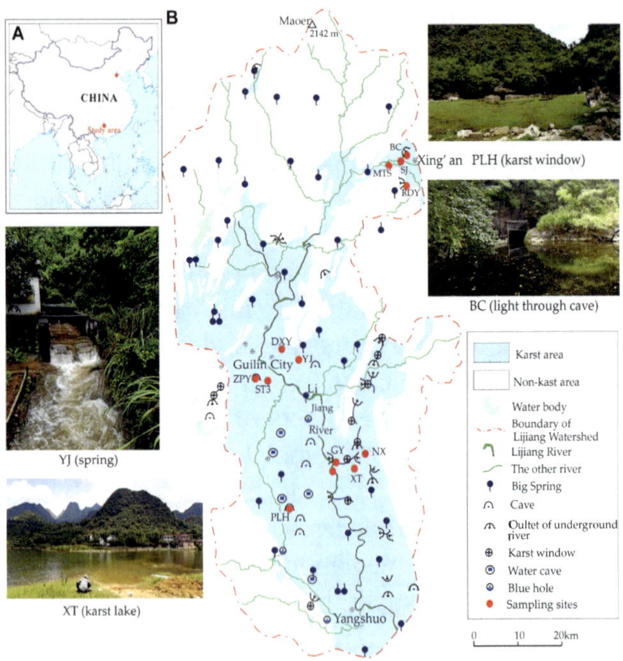

Figure 3.5. Wetlands.

Source: *Guo F, Jiang G and Liu F (2022) Plankton distribution patterns and the indicative significance of diverse cave wetlands in subtropical karst basin. Front. Environ. Sci. 10:970485. doi: 10.3389/fenvs.2022.970485*

The amazing ability of wetlands (Figure 3.5) to hold developing environmental toxins, notably heavy metals and organic pollutants, from a variety of industrial and agricultural effluents is due to the specific properties of wetlands that stimulate numerous biogeochemical processes. Interestingly, plants are essential elements of both natural and artificial wetlands in the clean-up of developing toxins. Wetlands possess a complex mechanism to remove the contaminants by sorption (adsorption or absorption). In this situation, phytoremediation, an eco-friendly phytotechnology, is used by hyperaccumulator plants, notably wetland plants, to rehabilitate the environment from numerous toxins. Mechanisms related to phytoremediation and biochemical changes are essential to phytoremediation technology for their increased efficiency in elucidating the destiny of emerging contaminants in aboveground and belowground plant sections of wetland plants. To achieve this, technological innovation, omics like proteomics, genomics, metabolomics, etc., and genetic engineering were used as unique tools to significantly increase the effectiveness of phytoremediation technology (Gerhardt, Gerwing, and Greenberg, 2017).

3.5 RHIZOSPHERE MICROBES

Phytoextraction, rhizofiltration, phytostabilization, phytodegradation, rhizodegradation, and phytovolatilization are some of the several ways that plants and the bacteria in their rhizosphere interact with contaminants. Pistia stratiots (Figure 3.6) and Eichhornia crassipes, (Figure 3.7) secrete exudates that supported bacterial strains and help in the removal of pollutants. Microbial relationships with wetland plants are intrinsically associated with phytoremediation success (Victor et al, 2016).

Figure 3.6. Pistia stratiots.

Source: *https://en.wikipedia.org/wiki/File:Pistia_stratiotes0.jpg*

Figure 3.7. Eichhornia crassipes

Source: *https://upload.wikimedia.org/wikipedia/commons/c/c3/Eichhornia_crassipes_C.jpg*

The phytostabilization process tends to protect the aboveground plant parts from hazardous contaminants by reducing the mobility and bioavailability of contaminants by their precipitation inside roots and the rhizosphere. However, only rooted wetland plants like Typha domingensis and Phragmites australis may use this approach (Bonanno, 2013). Additionally, phytotransformation and phytodegradation use enzymes or their cofactors to break down or immobilize pollutants in plant roots and shoots. The most important mechanism for polluted soil and water is phytoextraction; nevertheless, it should not be used on agricultural crops due to negative effects on human health as well as contamination in the food chain. Both the aboveground (shoot) and belowground (root) sections of plants are used in phytoextraction, which has the ability to transport and concentrate a wide variety of pollutants. Rhizofiltration works similarly to phytoextraction, except it only involves the root section of the plant, which lies below ground. Limited research has been done on the removal of emerging environmental contaminants like volatile organic carbons, pesticides, PCPs, organics, and pathogenic microbes and heavy metals with global wetland plants. The majority of research carried out globally has demonstrated the removal of BOD, chemical oxygen demand (COD), suspended solids, and nutrient. Furthermore, there is absolutely no concern elsewhere about the fate of NPs and the involvement of wetland plants in their phytoremediation. Plants offer a great potential for the phytoremediation of NPs, which has not been well studied, as a green tool for emerging pollutants.

3.6 ORGANICS AND THEIR DEGRADATION

While non-volatile organics are sequestered or degraded inside the plant through phytoextraction and phytodegradation, volatile organics can be discharged through the leaves and into the atmosphere through transpiration. The phytovolatilization process can break down organic pollutants such acetone, phenol, chlorinated benzenes, chlorine ethenes, and BTEX (benzene, toluene, ethylbenzene, and xylene). By microbial activity also known as phytostimulation or rhizodegradation, organic materials can be broken down within the rhizosphere. Emerging pollutants, such as polybrominated diphenyl ethers (PBDEs), were eliminated using TiO_2 photocatalysis and phytoremediation using the Phragmites australis (Figure 3.8).

Figure 3.8 Phragmites australis.

Source: *Machaka M, Khatib J, Baydoun S, Elkordi A, Assaad JJ. The Effect of Adding Phragmites australis Fibers on the Properties of Concrete. Buildings. 2022; 12(3):278. https://doi.org/10.3390/buildings12030278*

In particular, organic pollutants like pentachlorophenol, perchloroethylene (PCE), and trichloroethylene (TCE) have the potential to be detoxified by the unusual but well-known wetland plant waterweed (Eichhornia crassipes) (Rezania et al., 2019). Designing treatment wetlands in man-made wetlands may benefit from an investigation of the water quality for organic pollutants. Salt marshes or the viability of the natural wetland environment are hampered by oil or hydrocarbon contamination.

Figure 3.9 Juncus roemerianus.

Source: *https://upload.wikimedia.org/wikipedia/commons/3/3f/Juncus_roemerianus.jpg*

Juncus roemerianus (Figure 3.9) is a reliable phytoremediator for wetlands that have been contaminated by hydrocarbons, petroleum, or diesel. Other wetlands plants have been used for the phytoremediation of polycyclic aromatic hydrocarbons (PAHs), including Juncus effusus, Iris pseudacorus, Phalaris arundinacea, and a grass mixture made up of Festuca arundinacea, Festuca rubra, and Lolium perenne. The transformation of explosives containing emerging contaminants, such as 2,4,6-trinitrotoluene (TNT), by aquatic plants was observed using the common wetland species Myriophyllum spicatum and Myriophyllum aquaticum, and remediation for TNT and Research Department Explosive (RDX) took place in a constructed wetland (Pavlostathis et al, 1998). The removal of emerging contaminants like PAHs, which have been designated a priority pollutant by the U.S. Environmental Protection Agency (EPA), is accomplished using constructed wetland technology in its various forms with a good success rate.

In the production of polymers, vinyl chloride and polyacrylates, the synthetic substance bisphenol A (BPA) (2,2-bis[4-hydroxyphenyl]-propane) is utilized. The primary sources of BPA release into the environment are wastewater treatment facilities that treat industrial effluents, sewage sludge, and waste landfill leachates. The world produces 3 million tonnes of BPA annually. Juncus acutus, a helophyte found in wetlands, has proven to be very helpful for the clean-up of developing pollutants. BPA seriously harms human health by causing various chronic diseases, including malignancies, respiratory diseases, autoimmune disorders, reproductive issues, chronic renal diseases, birth defects, and developmental problems. Micropollutants are emerging contaminants in the entire environment that are typically detected in trace amounts, that is, between nanograms per liter and micrograms per liter, or even lower. Constructed wetlands were made possible by the failure of conventional techniques and sewage treatment facilities to effectively remove these environmental micropollutants.

3.7 PCP CONTAMINATION

It is interesting to note that personal care products (PCPs) are essential ingredients in nutraceuticals, fragrances, sunscreen agents, and commonly used medications like analgesics and anti-inflammatories, antidiabetics, antiestrogenics, antiprotozoals, antiseptics, lipid regulators, diuretics, medications for treating pulmonary arterial hypertension, psychiatric drugs and antidepressants, psychostimulants, etc. They are utilized in products like cosmetics and PCPs, surfactants, and phytosanitary items that eventually act as pollutants in the environment. Dietary consumption of these PCP-contaminated plants may seriously endanger human health in addition to disturbing environmental health and ecological change, but at the moment, little is known about the fate of PCPs in plants or their impact on or risk to the ecosystem. PCPs primarily enter plants through the use of reclaimed wastewater for irrigation, the application of biosolids and manure for fertilizing agricultural soils, and deposition from volatile compounds. Ecosystems in wetlands and other aquatic environments are particularly vulnerable to PCPs. The direct discharge of untreated PCPs into sewage should be prevented to reduce the harmful effect on the environment. Care may be needed to enhance the breakdown of PCPs to harmless compounds in waste treatment plants. Lemna gibba L. treated with ibuprofen was found to have no phytotoxic effects through growth and physiological measures, demonstrating that wetland plants may be tolerant to PCPs (Bokhari et al, 2016). Therefore, the most environmentally sound method of addressing these developing toxins is by phytoremediation of PCPs using wetland plants. Some Juncus species, like J. phaeocephalus and J. patens, worked with another wetland plant, Hydrocotyle spp., to remove organic pollutants like diazinon, dichlorodiphenyl-trichloroethane (DDT), total chlordanes, and total organochlorine pesticides (McKnight, Gannon, and Yelverton, 2022). Juncus species have the ability to sequester metals inside their aboveground as well as belowground biomass.

The growing pollutants of concern in this context are beta-blockers, cancer drugs, hormones, analgesics and anti-inflammatory medications, chemical compounds used for disinfection and cleaning, hormones, and endocrine-disrupting substances. The majority of rivers, lakes, and coastal waterways, as well as groundwater and some drinking water in populated regions, become contaminated because sewage treatment plants are unable to completely degrade these PCPs or their partially degraded intermediates. Agriculture systems receive untreated PCPs via irrigation water from treatment facilities, which could contaminate the plants that make up the majority of human dietary needs. Therefore, the expansion of this pollutant poses substantial threats to human health. Sewage sludge-infused soil additives make the issue worse and contaminate the soil. Additionally, spreading veterinary antibiotics through the application of animal dung on agricultural land contributes to their release into the environment. Phragmites australis was used in research for the phytoremediation of veterinary medicines, and it was found to remove 94 percent and 75 percent, respectively, of enrofloxacin and tetracycline from wastewater. Similar to this, the vetiver grass Chrysopogon zizanioides remedied tetracycline and accumulated it primarily in aboveground biomass (Chen et al, 2004). Typha, an emergent wetlands plant, showed that it could phytoremediate clofibric acid-contaminated waters. Phragmites australis, a common marsh plant, demonstrated the effectiveness of phytoremediation for the drugs ciprofloxacin, oxytetracycline, and sulfamethazine. The phytoremediation of tetracycline and oxytetracycline in aqueous media was studied using wetlands plants like Myriophyllum aquaticum and Pistia stratiotes, and root exudates consisting of enzymes and metabolites played a significant role in the degradation of these emerging contaminants of concern (Mustafa and Hayder, 2021).

Future options for the phytoremediation of developing pollutants, like fluoxetine and ibuprofen, are potentially offered by wetland plant-based reactors. From this prospect, the duckweed reactors (Landoltia punctata and Lemna minor) appears an excellent tool. When important and potent wetland plants like Salvinia molesta, L. minor, Ceratophyllum demersum, and Elodea canadensis interacted with various emerging contaminants under the category of PCPs in microcosms, it was discovered that caffeine was easily remedied by the free-floating plants L. minor and S. molesta, followed by E. canadensis, and C. demersum (Majeti Narasimha Vara Prasad, 2019). Similarly to this, several studies have shown that poplar cells completely removed the drug ibuprofen from the nutritional media. Recent studies have looked into the 96–99 percent removal efficiency of azole herbicides like tebuconazole and imazalil when remedied using the wetlands plant Phragmites australis.

Physical and chemical mechanisms used in the remediation of pathogenic microbes have limitations that have given way to biological or phytoremediation using wetland plants. The composition and design of the hydraulic regime, retention period, and density of wetland plants in man-made wetlands have a significant impact on the extent of pathogenic microorganism remediation. In artificial wetlands, the rhizospheres of rooted wetland plants contain a variety of exudates. Wetland plant roots, soil, and microbial relationships become complicated as a result of biotic and abiotic stressors in the rhizospheric environment. Exudate antimicrobial qualities have a negative impact on or eradicate harmful bacteria while retaining helpful microbes that improve the rhizospheric physical and chemical characteristics. When exposed to certain chemicals, some plants, like sweet basil (Ocimum basilicum), release exudates as a response. When these plants are exposed to Pseudomonas aeruginosa, they release antibiotic exudates like rosmaric acid (Sai Saraswathi et al, 2017).

While Sinorhizobium meliloti infestation causes Arabidopsis thaliana to release peroxidases as a form of defence against the pathogen, Pseudomonas syringae infestation causes the opposite reaction. These findings suggest that the release of exudates by wetland plants in their rhizospheric environment may be microbial pathogen-specific.

It has been noted that plant-secreted proteins such as lectins trigger the development of compact microbial aggregates on the root surface, increasing pathogen clearance from wastewater. The importance of marsh plants in the phytoremediation of harmful microorganisms has been shown in numerous international case studies. Water hyacinth wetlands are frequently employed for the treatment of various contaminants in artificial wetlands in order to avoid significant restrictions on land use. Sand filtration is a very thoughtful approach to get rid of emerging pollutants in general and especially the pathogenic microorganisms,.

3.8 NEGATIVE EFFECTS OF MEDICINES AND THEIR METABOLITES

Due to the negative effects that pharmaceuticals and antibiotics have on the environment and human health, they have been the subject of an increase in phytoremediation studies during the past ten years. Inadequate levels of medicines and their metabolites have unfortunately been introduced into the aquatic environment through sewage, which carries human excreta from those who have used these drugs, and agricultural runoff, which contains cattle manure. Wetland plants found in constructed wetlands could be a useful tool for eliminating medicines.

To this purpose, developing an understanding of their kinetics is essential to enhancing the removal of pharmaceuticals by artificial wetlands. In phytoremediation research on pharmaceutical compounds or antibiotics, Typha latifolia, Phragmites australis, Iris pseudocorus, Juncus effusus, and Berula erecta are powerful wetland plants in constructed wetlands. An integrated biochemical and proteomic analysis of Cyperus alternifolius interaction with pharmaceutically active compounds (PhACs) revealed molecular mechanisms in a constructed wetland (Yan et al, 2016). It was discovered that reactive oxygen species could be successfully combated by increased antioxidant enzyme activities, leading to the restoration of the photosynthetic pigments.

The use of living organisms to remove new pollutants has recently drawn a lot of public interest and research and development (R&D) funding. Because it is solar-powered and can be done on-site, using plants to remediate emerging pollutants is an appealing solution that reduces both cost and human exposure. Endophytes support plant growth by producing phytohormones, supplying nitrogen following nitrogen fixation, defending against environmental stresses (heat, cold, drought, and salt), and enhancing phytoremediation by enhancing contaminant uptake and toxin degradation. They can both degrade pollutants and are resistant to new contaminants. Bacteria have been identified through 16S rRNA gene sequencing in the context of textile effluents containing emerging contaminants and their phytoremediation. A number of procedures, including phytoextraction, rhizofiltration, phytostabilization, phytovolatilization, and phytotransformation or phytodegradation, make up the phytoremediation mechanism for emerging pollutants. Each phase plays a unique part in the buildup and removal of developing pollutants.

3.9 PHYTOEXTRACTION

Phytoextraction is the use of plants to remove new contaminants from bodies of water by accumulating them in their tissue, particularly in the harvestable roots and shoots. Various methods of phytomining may result in the recovery of metals from biomass. Compared to other plants and the ambient metal concentration, some hyperaccumulator plants absorb disproportionately high levels of developing pollutants, such as metals. Heavy metal liquefaction and buildup during the phytoextraction process result in relatively stable metal fractions, which reduce the bioavailability and ecotoxicity of heavy metals. Following multiple attempts, two phytoextraction techniques have emerged. Chelate-assisted phytoextraction, also known as induced phytoextraction, involves the use of synthetic chelates to improve the mobility and uptake of metal pollutants. Continuous phytoextraction, in which the plant's inherent capacity for remediation determines how much metal is removed. The only variable under control is the number of repeats of plant development. Rhizofiltration is the process by which macrophyte roots in aquatic bodies absorb emergent pollutants, which then precipitate and concentrate in their biomass. Rhizofiltration is a method used by wetland plants to accumulate and concentrate emerging pollutants or metals. Plants are grown hydroponically and then transplanted into waters that are contaminated with metals, where they absorb and concentrate the contaminants, particularly metals, in their roots and shoots.

Figure 3.10 Azolla pinnata

Source: *https://upload.wikimedia.org/wikipedia/commons/8/80/Azolla_caroliniana0.jpg*

The procedure entails chemisorption, complexation, ion exchange, microprecipitate formation, hydroxide condensation onto the bio surface, and surface adsorption. Utilizing plants with enhanced metal absorption and translocation capabilities can improve the efficiency of these procedures. Rhizofiltration is a particularly cost-competitive approach in the treatment of surface or groundwater containing relatively low amounts of hazardous metals employing wetland plants, such as Eichhornia crassipes, Lemna minor, and Azolla pinnata (Figure 3.10).

3.10 PHYTOSTABILIZATION

Phytostabilization is the immobilization of emerging contaminants by the use of tolerant wetland plants through absorption and accumulation in their tissues, adsorption in their roots, or precipitation within the root space, preventing the transfer of the contaminants in the soil as well as their movement by erosion. Instead of removing emerging toxins from a site, it aims to stabilize them through aggregation in root zones, reducing their mobility and avoiding migration to groundwater or the atmosphere while also lowering the risk to human health. The right choice of macrophyte is needed for the phytostabilization process. Characteristics of wetland plants include their ability to withstand high concentrations of contaminants of concern, their high root biomass production capacity to immobilize these contaminants through uptake, precipitation, or reduction, and their retention of applicable contaminants in roots rather than their transfer to shoots, which prevents the need for special handling and disposal of shoots. Phytovolatilization is the transpiration of pollutants that plants have absorbed and transformed into the atmosphere (Dhankher et al, 2012). By converting certain metals, like extremely hazardous mercury (Hg^{2+}) and methylmercury, utilizing the enzyme mercuric reductase to less dangerous forms, plants are able to volatilize these substances. The aerial components of the macrophyte, such as the leaves, are where the mechanisms and steps of phytovolatilization mostly take place. Phytodegradation is the removal or breakdown of metals by a macrophyte's enzyme or enzyme cofactors. Inside the plant bodies, enzymes including dehalogenases, oxygenases, and reductases break down molecules that include emerging pollutants. Rhizodegradation also takes place with the assistance of the related bacteria in the plant's root zone.

The enzyme phosphatase, found in marsh plants like gigantic duckweed (Spirodela polyrhiza), is utilized to dissolve the chemical connection between phosphate and organophosphate insecticides (Appenroth et al, 2010). Wetland plants including parrot feather (Myriophyllum aquaticum), stonewort (Nitella spp.), and hybrid poplar (Populus spp.) have the enzyme nitroreductase, which tends to reduce nitro groups on explosives and other nitroaromatic chemicals and remove nitrogen from ring structures. Wetland plants like willow (Salix spp.) contain nitrilase, which is utilized to separate cyanide groups from aromatic rings. Stonewort and parrot feathers (Myriophyllum aquaticum) are wetlands plants that contain laccase, which is needed to break aromatic rings once TNT is converted to triaminotoluene. Dehalogenase, which tends to dehalogenate chlorinated solvents, is found in wetlands plants including hybrid poplar (Populus spp.) and several algal species, including parrot feather (Myriophyllum aquaticum). Horseradish (Armoracia rusticana), a distinctive wetland plant, has the peroxidase enzyme. For the successful phytoremediation of emerging pollutants in contaminated water bodies, many types of wetland plants are utilized. In order to stop additional metal contamination of the water bodies, these aquatic wetland plants and the bacteria they are linked with were used to absorb and break down the metals. , It is necessary to cooperate, integrate, and assimilate these biotechnological advancements with traditional and ethical understanding to fully understand the obscurities of nature in the developing fields of bioremediation and phytoremediation.

Wetlands are a desirable choice for satisfying water quality regulations through nitrogen reduction for farmers and small industries because they provide additional economic benefits like vegetation for animal feed and habitat for harvestable fish. Free-floating macrophytes shade the water column, allowing fish and macroinvertebrates to live in a cooler environment in what would otherwise be a warm-water tropical climate. Although wetlands with floating aquatic plants support larger mosquito larvae populations than open-water areas due to lower dissolved oxygen concentrations, floating aquatic plants do provide cooler water temperatures and abundant food sources that help create an ideal habitat structure for fish and invertebrates. Due to its rapid growth rate and substantial intake of nutrients and pollutants, water hyacinth (Eichhornia crassipes) is one of the most frequently employed plants in artificial wetlands. Its tendency to bioaccumulate and biomagnify the heavy metal pollutants prevalent in water bodies has been investigated. Metals are accumulated, and the recycling process is cost-effective and sustainable since it is powered by photosynthesis and biomass growth. It can store metals like Cd, Cu, Pb, and Zn in the tissues of its roots. Due to its exotic invasive character and quick decomposition when compared to other plants, it has been observed that the spread of water hyacinth provides a challenge for the operation of built wetlands.

3.11 WASTEWATER CLEANING

Wastewater is cleaned of Cr by macrophytes and wetland plants such Eichhornia crassipes, Pistia stratiotes, and Spirodela polyrrhiza. As they are cold tolerant than water hyacinth, easier to harvest than algae, and capable of rapid growth, duckweeds (family Lemnaceae) appear to be the superior alternative and have been suggested for industrial effluent treatment (Mishra & Tripathi, 2008). By filtering out heavy metals, the biomass of aquatic wetland plants, like Azolla, can be utilized to recycle municipal wastewater for irrigation, helping to save water resources. In addition, wetland plant waste biomass that is created after treatment can be used to produce biogas in an effort to achieve sustainability in the energy industry. Due to its interaction with the cyanobacterium Anabena azollae, Azolla pinnata may have enormous biofertilizer potential after being treated with HCl or HNO_3 to eliminate heavy metals absorbed. Typha species function as nutrient pumps, drawing substantial amounts of nutrients from the substrate and storing them in the tissue above ground. Despite the increased Cd uptake, they showed no discernible symptoms of toxicity. Phalaris arundinacea, also known as reed canary grass, is a powerful bioresource for the removal of trace metals from bottom sediments that certain metals, such as Zn, Fe, Pb, Cu, Ni, Cd, and Cr, concentrate in larger amounts inside the root, followed by the leaf and stems. Eleocharis acicularis roots and shoots successfully phytoremediated indium, Ag, Pb, Cu, Cd, and Zn (Khan & Faisal, 2018).

Metals may also precipitate on root surfaces as a result of changes in the pH of the rhizosphere and root exudates. Roots or whole plants are picked for disposal as they become saturated with metal contaminants or other developing contaminants. Due to their capacity to extract up to 60% of their dry weight as harmful metals, rhizofiltration using plants significantly lowers the cost of producing and disposing of the hazardous residue. In order to treat surface or groundwater with relatively low amounts of harmful metals, rhizofiltration is a very cost-competitive approach. Plants used in phytoremediation should only accumulate new pollutants in their roots. The efficacy of rhizofiltration would be reduced by the increased amount of contaminated plant residue that would need to be disposed of. Process efficiency may be improved by utilizing plants with a greater capacity for metal absorption and translocation. In addition to being simple to handle, inexpensive to maintain, and requiring the least amount of secondary waste management, plants should be able to accumulate and tolerate considerable concentrations of the target metals. It is ideal for plants to produce a lot of root biomass or root surface area. Chemical industries, such as chlor-alkali facilities in developing nations like India, as well as coal mining and its associated sectors, such as thermal power plants, are sources of growing pollutants like heavy metals. Rivers, lakes, and reservoirs' water quality and aquatic biodiversity are seriously threatened by the effluents from these enterprises. At all levels, the issue of developing pollutants and heavy metal pollution is becoming more concrete. Wetlands have a very high ecosystem service value, and heavy metal and developing pollutant pollution in the aquatic ecosystem pose a serious danger to aquatic biodiversity and serious health risks for humans. Fisheries, irrigation, residential water supply, recreation, and tourism have long been supported by wetlands.

Peatlands, mangroves, marshes, and swamps are examples of wetlands that contribute to the carbon cycle. Long-term carbon storage occurs in wetland sediments, while short-term storage occurs in dissolved substances in surface and groundwater and the wetland biomass of plants, animals, bacteria, and fungus. Since they may absorb significant quantities of environmental pollutants, natural wetlands have been employed for centuries as a waste disposal system. Wetlands are a low-cost method of reducing pollution from both point and nonpoint sources. Wetlands also serve as the "kidneys of nature," capable of effectively modifying a wide range of pollutants. These ecosystems have unique qualities that make them especially well-suited for wastewater treatment: they are semi-aquatic systems that typically contain large amounts of water, they have oxic and partly anoxic soils where the biodegradation of organic matter occurs, and they support highly productive, tall emergent vegetation that is capable of absorbing a large amount of nutrients that promote growth. They carry out numerous ecological roles and support wildlife, which may include a number of uncommon, threatened, and endangered species. Due to urbanization, which has contributed to rising pollution and contamination, wetlands are one of the habitats that are rapidly degrading. A heavy strain has been placed on the ability of aquatic ecosystems to self-clean as a result of the relatively recent arrival of anthropogenic-origin harmful compounds, particularly heavy metals, and their enormous transfer to diverse environmental compartments, especially water. Heavy metals are extremely harmful to living things, notably humans. In order to achieve sustainability and use a green strategy to remove pollution, phytotechnology is essential. The issue of heavy metals has gotten worse due to growing industry and urbanization.

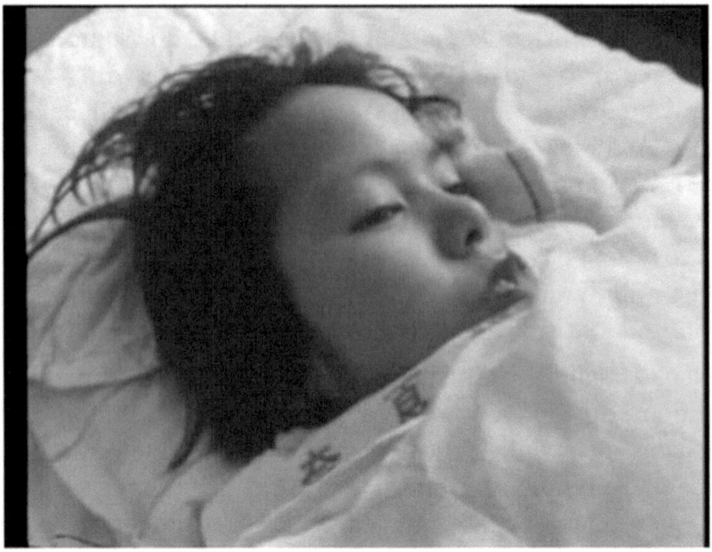

Figure 3.11 Minamata disease patient Kumiko-The medium shot of Kumiko awake in bed. Still from The Children of Minamata are Living (1965), director Tsuchimoto Noriaki.

Source: *Inoue M. The Ethics of Representation in Light of Minamata Disease: Tsuchimoto Noriaki and His Minamata Documentaries. Arts. 2019; 8(1):37. https://doi.org/10.3390/arts8010037*

3.12 POLLUTION

Since the start of the Industrial Revolution, hazardous metal pollution of the biosphere has increased significantly. At a local, regional, and international level, heavy metal contamination is becoming a problem. Over the previous few decades, increased mercury contamination from a chemical company in Japan caused Minamata disease, (Figure 3.11) (which impacted 2252 people and claimed 1043 lives. Large-scale issues have recently emerged in developing nations, such as contamination of groundwater in Bangladesh and contamination of drinking water sources with Cd, Pb, Cu, and Zn in Bolivia, Hong Kong, and Berlin (World Health Organization. Regional Office for Europe, 2021). Both natural and man-made sources of heavy metals can be found in surface water systems. Metal imports from humans currently outweigh those from natural sources. In aquatic and terrestrial ecosystems, high concentrations of Cd, Cu, Pb, and Fe can serve as ecological toxins. The ecosystem and people may both be at risk from surface water with excessive metal levels. Due to runoff from fertilized land, agriculture appears to be the largest source of pollution from emerging pollutants. The primary anthropogenic sources of emerging pollutants and metal contamination are thought to be industrial processing and solid waste dumps. Metal pollution and human history are closely related as a result of ancient mining and metalworking. Municipal waste, fertilizers, pesticides, and sewage are the main sources of pollutants and metal contamination. Where waste materials containing metal-rich sulfides from mining activity have been stored or abandoned, emerging pollutants or heavy metal contamination as well as acid mine drainage (AMD) are very relevant concerns (Thomas, Sheridan, & Holm, 2022). Coal mining and its related industries, such as thermal power plants, as well as chemical industries, such as chloralkali plants, are additional sources of emerging contaminants. These industries are major sources of heavy metals in industrial belts in developing nations like India.

Gold (Au) and silver (Ag) mines, the coal industry, untreated abandoned batteries, and industrial waste disposal are major sources of pollutants, such as mercury. Electronic components, electroplating, metal processing, and mining are some industries that use emerging pollutants like Cd. Plastic toys and food containers are other places you can find them. Sewage sludge, mining, and mineral processing are examples of anthropogenic sources of manganese in the environment. Emerging pollutants and heavy metals are fast becoming one of the biggest dangers to wetland ecosystems. Worldwide interest has been focused on various emerging contaminants that have the potential to be toxic materials and environmental pollutants. The aquatic ecosystem being poisoned by heavy metals and developing pollutants has been a pressing issue everywhere. Aluminum (Al), arsenic (As), cadmium (Cd), copper (Cu), cobalt (Co), chromium (Cr), mercury (Hg), iron (Fe), lead (Pb), magnesium (Mg), manganese (Mn), nickel (Ni), strontium (Sr), vanadium (V), and zinc are some of the rising pollutants and heavy metals (Zn). These heavy metals are the main cause of aquatic contamination and are persistent harmful compounds. In addition to having a toxic effect on biological systems through bioaccumulation and biomagnification, heavy metals and emerging contaminants can also have a hazardous impact on the environment of wetlands and the associated living organisms. Since they are crucial to both human health and the health of the environment, their toxicity has been a major concern. When harmful emergent pollutants are found in lakes, reservoirs, and rivers, it has an impact on the lives of the locals who depend on these water sources for their daily needs.

Due to their chemical, physical, and biological processes that include adsorption, precipitation, sedimentation, and induced biogeochemical changes by plants and bacteria, wetlands have long been recognized as an important emerging contaminant and a heavy metal sink. Human activities and the release of new contaminants, such as persistent organic pollutants and heavy metals, are degrading the wetlands. The metabolism, growth, and development of plants are restricted by an excessive concentration of developing pollutants in soil solutions, such as Co, Cu, Fe, Mn, Mo, Ni, and Zn. Genetic diversity in many microbes enables them to avoid the toxic effects of new pollutants, while more advanced species typically succumb to the toxic consequences. More particular, harmful emerging pollutants, including methylmercury consumption from fish and other aquatic foods, can significantly impact human health. Higher concentrations of mercury can harm important organs including the lungs and kidneys. Methylmercury may get through the placental barrier and affect the developing brain. Renal dysfunction and bone diseases, like itai-itai in Japan, can result from the buildup of new pollutants like Cd in the human body. Gallbladder carcinoma and stone formation have been linked to high cadmium concentrations in sewage, irrigation water, and vegetables grown in the Gangetic Plain of eastern Uttar Pradesh and western Bihar, with cadmium, chromium, and lead concentrations being significantly higher in gallbladder carcinoma than in gallstones. Children who are lead poisoned experience neurological damage that lowers their IQ, impairs their short-term memory, causes learning difficulties, and affects their coordination. Arsenic side effects include peripheral neuropathy, renal damage, skin cancer and other skin conditions.

3.13 HEAVY METALS

Pb, Cr, and As are among the dangerous heavy metals that can seriously harm both people and animals. They can cause inflammation, respiratory problems, arsenic poisoning, and damage to the sensory organs. Emerging contaminants in drinking water, like As and Cd, can harm the bones, neurological system, kidney, liver, digestive system, kidney, and skin, and cause hyperpigmentation, allergies, and cancer. Long-term exposure to low levels of cancer-causing heavy metals and new pollutants may cause a number of malignancies. Cd is an inorganic heavy metal that is becoming more and more carcinogenic, and it has a 38-year half-life, which means that it can exist in the body for a long time. Because of its interactions with biological systems, Cd can enter living organisms through the respiratory system and cause damage to tissues and organs upon contact with the skin. The possibility that harmful pollutants, including heavy metals, could accumulate in animal bodies or harm human DNA through the food chain is another thing increasing doubts about these contaminants. They have carcinogenic effects as a result of their ability to mutate. For instance, specific species of Cd, Cr, and Cu have been linked to adverse health effects, ranging from dermatitis to various malignancies. Nevertheless, the possibility of these toxins leaking into the environment following phytoremediation raises environmental concerns. Consequently, uncertainties about the long-term fate and secure disposal of these contaminated biomasses confuse phytotechnologists. Water hyacinth, for example, is an invasive species that requires sustainable management. Therefore, even after phytoremediation, NPs and developing pollutants are not entirely removed from the environment. Incineration, composting, and landfilling are some of the currently used conventional methods for the removal of hyperaccumulator plants, but these techniques have the tendency to recycle these organic or inorganic contaminants into the environment, causing secondary pollution. The produced biomass, which is free of environmental toxins, can be reused. Wetland macrophyte bioresources should be promoted for use in phytoremediation technology to replace conventional, expensive, and inefficient technologies. It is ideal for a developing country with severe water pollution, like India, particularly in wetlands. Another key idea for making use of readily accessible plant resources is the exploitation of invasive plant species like Eichhornia crassipes. The removal of harmful heavy metals and other contaminants from wetlands using wetland bioresources will assist the nearby living things by giving them a safe habitat free from toxins. Green phytotechnological innovation might benefit from a concerted multidisciplinary strategy integrating macrophyte bioresources with wetland ecology, pollution science, environmental engineering, genetic engineering, and sustainability concerns. Microbial productivity, which is integrally tied to wetland plants, can be improved for developing phytotechnologies, and it may have a promising future.

Due to the ongoing destruction of natural wetlands, numerous artificial wetlands had to be built in order to serve as treatment wetlands for toxins that were developing as a result of various human activities (Macek et al., 2004). In a natural wetland, plants offer a substrate (roots, stems, and leaves) for microbes to develop as they consume heavy metals and break down various developing pollutants, such as organic materials. However, it has been acknowledged that natural wetland ecosystems do not always function effectively for desired objectives, such as the removal of recalcitrant emerging contaminants and the maintenance of strict water quality standards, as a result of the exponentially increasing demands of human expansion and resource exploitation. A built wetland is an artificial marsh or swamp that has been created to use wetland vegetation, soils, and their associated microbial communities in natural processes to help clean waste and various developing contaminants.

In order to remediate emergent pollutants that are present in a water environment, constructed wetlands and treatment wetlands are engineering systems that have been created to use natural processes, such as plant, soil, and microbial activity. Additionally, constructed wetlands have the advantages of being inexpensive and low-maintenance and are capable of removing a variety of emerging contaminants, such as micropollutants, heavy metals, nutrients, organic matter, and pharmaceuticals and personal care products (PCPs). Additionally, in recent years, constructed wetlands have been used to treat a variety of emerging contaminants found in different types of wastewaters and effluent, such as landfill leachate, nonpoint-source stormwater runoff, agricultural wastewater, acid mine drainage, point-source domestic sewage, and agricultural wastewater. The majority of emergent pollutants are removed from wastewater in artificial wetlands via substrate interactions, with plants acting as a "polishing system". The anoxic soils contribute to the conditions necessary for emerging pollutants, such as heavy metals, to be immobilized in their highly reduced sulfite or metallic forms. Meanwhile, plants are crucial for the removal of metals through filtration, adsorption, and cation exchange, as well as through plant-induced chemical changes in the rhizosphere.

3.14 ROOT ZONE PHYTOREMOVAL

The root zone is where man-made wetlands react most quickly. The interplay of plants, microbes, soil, and pollutants causes these physicochemical and biological processes for the phytoremoval of many toxins. In many ways, wetland phytoremediation is quite different from terrestrial phytoremediation. The categories of discharges or effluents that may only be treated through wetland systems are domestic wastewater, agricultural runoff, urban storm runoff, acid mine drainage, and microbiological pathogens. Wetland biosystems are better able to handle industrial effluent than terrestrial phytoremediation. The anoxic soils that typically characterize constructed wetlands have a tendency to immobilize pollutants, which is noticeably different from terrestrial phytoremediation, where the oxidized soils tend to mobilize pollutants into plant tissue. This is particularly true of emerging contaminants like heavy metals and some organics. Furthermore, because they are made for certain toxins and are better adapted to remediate them within strict discharge limitations, engineered wetlands may be more effective than natural wetlands. While the treatment of developing pollutants using traditional technologies can be an energy-intensive process, solar energy is particularly cost-effectively used by wetland plants in artificial wetlands.

3.15 ARTIFICIAL WETLANDS

Large-scale artificial wetlands can successfully mimic natural wetlands' ecosystem functions and speed up the restoration process, yet research on how well ecosystem structures restore in terms of living biomass and water-using energy value is still preliminary (Williams, 2002). When compared to more sophisticated chemical technologies, phytoremediation and its application in artificial wetlands are among the more efficient and affordable eco-technologies. These phytotechnologies encourage the safe and effective use of macrophytes and wetlands plants in medicine. The technology of artificial wetlands can be used to treat a variety of developing toxins from polluted water. By doing this, emerging pollutant concentrations are reduced. Over the course of six years, it has been shown that constructed wetlands may remove a sizable portion of developing contaminants from road runoff, such as the heavy metals Cd, Cu, Pb, and Zn. All contaminated aquatic bodies, including rivers, ponds, lakes, and estuaries, can use this method. Nowadays, artificial wetlands are designed with selective macrophytes for efficient phytoremediation processes.

Increased concentrations of a number of developing toxins are the result of the excessive load created by a growing population and rapid industrialization. Therefore, the phytoremediation of emerging pollutants in natural wetlands did not succeed just by relying on wetlands. Furthermore, natural wetlands faced fewer restrictions than artificially created ones due to seasonal, environmental, and climatic reasons. Due to these elements, as well as the variety of new pollutants and harmful bacteria, natural wetlands served as a model for artificial wetlands. Due to some existing flow, pollutant-loaded water has a limited retention time in natural wetlands. In addition, the yearly mass balance for nutrients in natural wetlands frequently demonstrates that seasonal impacts are also a limitation. Emerging pollutants that are toxic and resistant can harm the health of the wetland ecosystem by building up in the plants there and killing off reliant wildlife, such as fish and birds. Interestingly, the separation of the water regime from natural patterns is the most striking difference between artificial and natural wetlands. Due to sudden changes in the water regime, the population dynamics of wetland plants are altered seasonally in natural wetlands. Man-made habitats that resemble natural wetlands are called constructed wetlands. The low energy requirements, simple operations, and little maintenance requirements of natural wetlands are their most appealing benefits. Following observations of the significance of natural wetlands, artificial wetlands are created to carefully accommodate human demands. Through the use of diverse carefully chosen materials, they are created to serve more purposes than actual wetlands. The majority of CWSs are used in the phytoremediation process, which involves picking the right wetlands plant, to clean up polluted water. They are developed from a non-wetland ecosystem or a previous terrestrial setting, mostly for the aim of removing pollutants or toxins from wastewater, as well as some pollutants, such as pharmaceuticals.

CHAPTER 4

Environmental Pollutants

CONTENTS

4.1 INTRODUCTION

There has been an enormous amount of toxic waste produced due to global industrialization, urbanization, and population growth. Numerous organic and inorganic substances included in this hazardous waste pose major risks to ecosystems. Chemicals such as petroleum hydrocarbons such as benzopyrene, chlorinated solvents such as polychlorinated benzenes, or PCBs, linear halogenated hydrocarbons such as trichloroethylene, volatile organic carbons, and explosives like trinitrotoluene are examples of organic pollutants. Nitrates, phosphates, metals and metalloids like uranium (U), strontium (Sr), and cesium (Cs), as well as non-radioactive or radioactive nuclides like mercury (Hg), arsenic (As), lead (Pb), cadmium (Cd), nickel (Ni), selenium (Se), silver (Ag), and zinc (Zn), are examples of inorganic compounds (Cs). These wastes are typically dumped on land or released into water bodies without sufficient treatment, despite pollution control measures, and become a significant source of environmental pollution and health risks. Therefore, it is crucial to remove these contaminants from polluted sites. Major environmental concerns currently arise from soil and water contamination. For decades, a variety of techniques based on physical, chemical, and biological methods have been available for the remediation of soil and water, but environmental protection calls for the development of more sustainable strategies that promise thorough, affordable, and environmentally friendly alternatives to current techniques. The removal of harmful pollutants from contaminated environments has recently been made more affordable, long-lasting, and aesthetically pleasing through phytoremediation. Plants were suggested for use in wastewater treatment over 300 years ago. The term phytoremediation refers to a wide range of plant-based green technologies that use either naturally occurring or genetically engineered plants to remove, transfer, stabilize, and/or degrade contaminants found in soil, sediment, and water. Natural habitats for a variety of microorganisms that can degrade toxins and improve treatment are provided by roots, stems, and leaves. Plants can remove contaminants from the environment through a variety of natural biophysical and biochemical mechanisms, including adsorption, transport and translocation, hyperaccumulation or transformation, and mineralization (Meagher, 2000).

4.2 POLLUTANT REMEDIATION

In order to remediate organic and inorganic pollutants from contaminated soil, sludge, sediment, groundwater, surface water, and wastewater, phytoremediation technology can be divided into a variety of categories, including phytoextraction, phytofiltration, phytostabilization, phytovolatilization, and phytodegradation, depending on the contaminants, the site conditions, the level of clean-up necessary, and the type of plants. The process of using plants to absorb, move, and store harmful pollutants from soil, sediments, and/or sludge in the root and shoot tissues is known as phytoextraction. It is also known as phytoaccumulation, phytoabsorption, or phytosequestration. The most popular and widely used phytoremediation method for removing hazardous metals from polluted environments is phytoextraction. Long-term care, regular plant harvesting, secure plant waste disposal, and maintenance of the plants are all necessary for phytoextraction. When compared to the costs of traditional soil restoration methods, phytoextraction is ten times less expensive. Thus, phytoextraction is a technology that is both efficient and affordable. Pb is one metal that phytoextraction cannot remove, but it can remove As, Cd, and Ni. In the root zone, phytoextraction takes place.

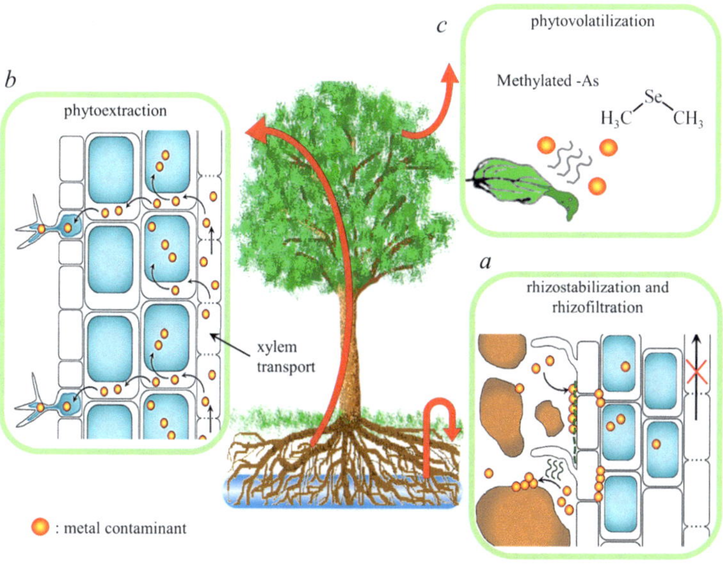

Figure 4.1. Continuous phytoextraction.

Source: *DalCorso G, Fasani E, Manara A, Visioli G, Furini A. Heavy Metal Pollutions: State of the Art and Innovation in Phytoremediation. International Journal of Molecular Sciences. 2019; 20(14):3412. https://doi.org/10.3390/ijms20143412*

The majority of the root may be located at a shallower depth as opposed to a deeper one, with the root zone being normally quite shallow. The process of remediating a contaminated site can also be impacted by a number of variables, such as the pH of wastewater and sediment, mobilization and uptake from the soil, compartmentalization and sequestration within the root, the effectiveness of xylem absorption and transport, distribution between metal sinks in the aerial parts, sequestration and storage in leaf cells, as well as plant growth and transpiration rates.

Continuous phytoextraction (Figure 4.1) is the term used to describe phytoextraction carried out using hyperaccumulator plants (Lombi et al, 2001). Continuous phytoextraction often relies on a plant's inherent capacity to absorb, translocate, and withstand high concentrations of metals during their whole growth cycle.

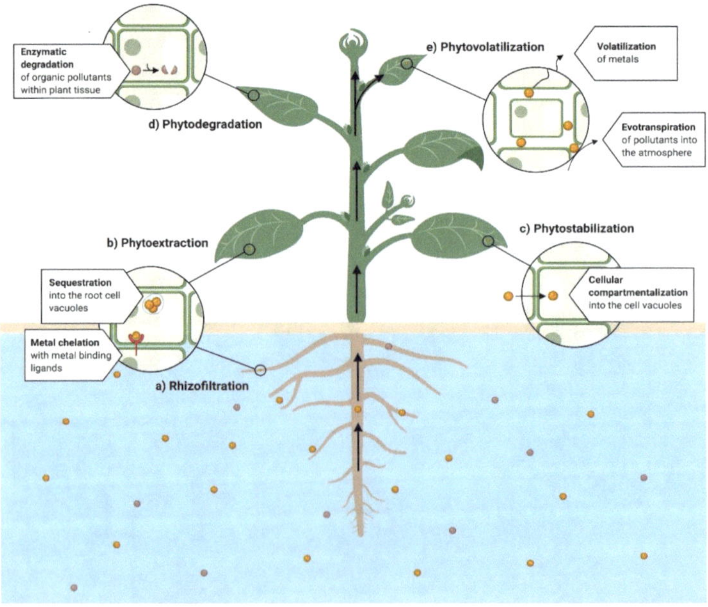

Figure 4.2. Rhizofiltration.

Source: *Delgado-González CR, Madariaga-Navarrete A, Fernández-Cortés JM, Islas-Pelcastre M, Oza G, Iqbal HMN, Sharma A. Advances and Applications of Water Phytoremediation: A Potential Biotechnological Approach for the Treatment of Heavy Metals from Contaminated Water. International Journal of Environmental Research and Public Health. 2021; 18(10):5215. https://doi.org/10.3390/ijerph18105215*

Continuous phytoextraction is also eco-friendly because it prepares the land for the growth of new plants. Continuous phytoextraction may be applied in two different types of sites in urbanized regions. In numerous research facilities across the world, the phytoextraction capabilities of ornamental plant species that are most frequently cultivated in urban settings are being studied.

4.3 RHIZOFILTRATION

With the use of the root zone technique known as rhizofiltration, (Figure 4.2) contaminants found in wastewater are interacted with by both terrestrial and aquatic plants. Contaminant mobility is decreased and their groundwater migration is prevented, which lowers their bioavailability and prevents them from entering the food chain. Metals absorbed only by the roots can be removed using rhizofiltration. When low pollutant concentrations and a considerable amount of water are involved, it is especially efficient and economically compelling. A plant that can extract hazardous metals from solution over an extended period of time and grow quickly is excellent for rhizofiltration. Rhizofiltration has a number of benefits, including the ability to be employed for in situ or *ex-situ* applications and the usage of other species besides hyperaccumulators. The maintenance of productive hydroponic systems in the field and the production of transplants produced hydroponically both require skilled staff, and the facilities and specialized equipment needed can raise overhead expenses. Adsorption, precipitation, rhizodegradation, and bioaccumulation are just a few of the physical and biological mechanisms that rhizofiltration uses to function. Rhizofiltration of metals refers to the metals present in complex industrial wastewater being absorbed into, adsorbing to, or precipitating onto plant roots. Rhizofiltration procedures could involve phytoextraction, phytostabilization, or phytovolatilization depending on the types of contaminants found in the wastewater. The effectiveness of the rhizofiltration process depends on specific compounds that lead to heavy metal accumulation in the plant body and subsequent precipitation on the root surface.

4.4 HEAVY METALS AND RHIZOFILTRATION

For their ability to remove heavy metals from contaminated streams through phytoremediation, several aquatic species have been discovered and evaluated. These species include *Oenathe javanica, Lepironia articulate, Hydrocotyle umbellate L., Lemna minor L., Eichhornia crassipes, P.*

stratiotes, and Polygonum amphibium L. However, due to their thin, slowly expanding roots, these plants have a limited capacity for rhizofiltration because they are ineffective at extracting metals. Indian mustard (*Brassica juncea*) and sunflower (*Helianthus annus L.*) are the most promising choices for removing metals from contaminated water (Rai, 2009). Sunflower plant can remove Pb, U, and Sr from hydroponic solutions, while the roots of Indian mustard have to be effective at removing Cd, Cr, Cu, Ni, Pb, and Zn as well. Through root adsorption and accumulation, adsorption on roots, or precipitation within the root zone, certain plant species encapsulate toxins in the soil during this process, lowering the risk to human health and the environment. The phytostabilization method may effectively immobilize a variety of heavy metals, including Pb, As, Cd, Cr, Cu, and Zn. In places with minute vegetation as a result of high pollution levels, phytostabilization can also be utilized to re-establish vegetation cover.

These rhizosphere-dwelling microbes like bacteria and mycorrhiza, which can actively contribute to changing the trace elements as well as help the plants, also play a significant part in these processes. The procedure is used when there are possible risks to human health and contaminants can lower exposure to dangerous compounds to safe levels. Since some organic pollutants and their metabolic by-products can adhere to or be integrated into plant components like lignins, there may be a chance for phytostabilization of organic pollutants. Phytolignification is the name given to this type of phytostabilization. However, one distinction between the two approaches is that, although phytostabilization of organic pollutants by phytolignification takes place aboveground, phytostabilization of metals is often meant to occur in the soil (Freitas, 2005). Contrary to phytoextraction or the excessive accumulation of metals in the shoots or roots, phytostabilization largely focuses on the sequestration of the metals in the rhizosphere rather than in plant tissues. Liming agents, phosphates (H_3PO_4), triple calcium phosphate, hydroxyapatite, phosphate rock, trace element (Fe/Mn) oxyhydroxides, organic materials such as biosolids, sludge, and composts, natural and synthetic zeolites, cyclonic and fly ash are often used as amendments to stabilize trace elements in contaminated soils. Biofertilizers decreased the uptake of As, Cr, and Zn by plants, whereas the application of organic amendments, such as dairy sludge, enhanced plant growth and decreased the bioavailability of As, Cr, and Zn in soil. In soils with Cr contamination, *Jatropha curcas* absorbs Cr via its roots and shoots. *J. curcas* experiences oxidative stress due to the accumulation of Cr but is able to cope with it due to overactive antioxidant defence mechanisms.

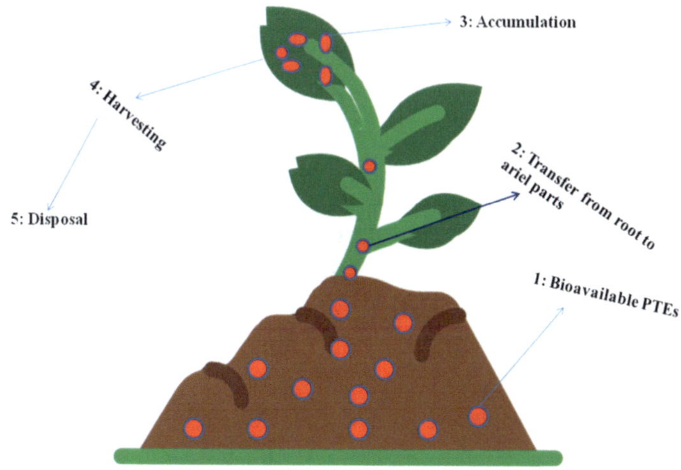

Figure 4.3. Phytovolatilization.

Source: *Wani ZA, Ahmad Z, Asgher M, Bhat JA, Sharma M, Kumar A, Sharma V, Kumar A, Pant S, Lukatkin AS, et al. Phytoremediation of Potentially Toxic Elements: Role, Status and Concerns. Plants. 2023; 12(3):429. https://doi.org/10.3390/plants12030429*

4.5 PHYTOVOLATILIZATION

Phytovolatilization (Figure 4.3) is the term used to describe the emission of volatile pollutants into the atmosphere by plant transpiration. A specialized type of phytotransformation known as phytovolatilization is only appropriate for pollutants with a high level of volatile nature. The biophysical mechanisms of mass movement through vegetation drive this intricate process. Once absorbed by plant roots, contaminants like Hg, As, and Se can be transformed into harmless forms and volatilized into the atmosphere from the roots, shoots, or leaves of the plant. Plants metabolize different inorganic or organic species of Se such as selenate, selenite, and Se-Met into a gaseous form during the process of Se phytovolatilization. The main process of Se removal from soil was proposed to be Se volatilization in the form of methyl selenate. Indian mustard (*Brassica juncea*) is a promising species for Se clean-up since it has a high rate of Se accumulation and volatilization as well as a rapid growth rate. The process of phytodegradation, often referred to as phytotransformation, occurs when pollutants that plants absorb are broken down either by internal metabolic processes or by the action of enzymes produced by plants on pollutants present nearby the plant.

Enzymes that catalyze and expedite the breakdown of organic contaminants can be produced by plants. Consequently, metabolic mechanisms break down organic contaminants into more manageable molecular forms, which are then absorbed into plant tissues to promote plant growth. Direct pollutant uptake from the medium and accumulation in the vegetation are prerequisites for phytotransformation-based site remediation. Direct chemical uptake into plant tissue through the root system is influenced by uptake effectiveness, transpiration rate, and chemical concentration in the soil and water. While transpiration rate is influenced by plant parameters such as type, leaf area, nutrients, soil moisture, temperature, wind speed, and relative humidity, absorption efficiency is dependent on chemical speciation, physico-chemical properties, and plant features. Herbicides and chlorinated solvents such as trichloroethylene can be broken down and converted by specific plant enzymes. This method can be applied to clean up fuel spills, landfill leachates, agricultural chemicals, and petrochemical sites and storage areas. The altered molecules that accumulate inside the plant must be either harmless or much less hazardous than the parent compound for this technology to be successfully implemented.

4.6 RHIZODEGRADATION

Rhizodegradation is a process in which plant-supplied substrates encourage the growth of microbial communities in the rhizosphere to break down organic pollutants in the soil (Dzantor, 2007). It is also known as phytostimulation, plant-assisted bioremediation, microbe-assisted phytoremediation, plant-aided in situ biodegradation, and enhanced rhizosphere biodegradation. The rhizosphere of a plant offers a special habitat for microorganisms that can convert toxic contaminants into by-products through metabolic activity. The biodegradation process in rhizodegradation is triggered by nutrients that the plant releases as root exudates. These exudates can lead to an increase in the microbial population and activity in the rhizosphere, which can boost the breakdown of organic contaminants in the soil. Rhizospheric microbes support plant growth by fixing nitrogen, mobilizing nutrients such as phosphorus, producing plant growth regulators, lowering stress hormone levels in plants, protecting them from infections, and degrading contaminants before they harm the plant. Rhizosphere bacteria frequently play a major role in the breakdown of contaminants. Due to environmental constraints at field sites like competition from weed species that are better adapted to the site, constrained plant growth in heavily contaminated soil, presence of plant pathogens, and other biotic and abiotic stressors, even enhanced rhizoremediation may be considerably slower than ex-situ treatments. Rhizoremediation is also ineffective for deeper layers and only works in the rooting zone. Because some harmful contaminant metabolites can bioaccumulate in plants, there must be precise guidelines on how to dispose of plant waste. Rhizodegradation has the advantage of destroying pollutants in their natural environments and has the capacity to completely mineralize organic contaminants. In addition, it has a lower likelihood of compound transfer to plants or the atmosphere than other phytoremediation techniques.

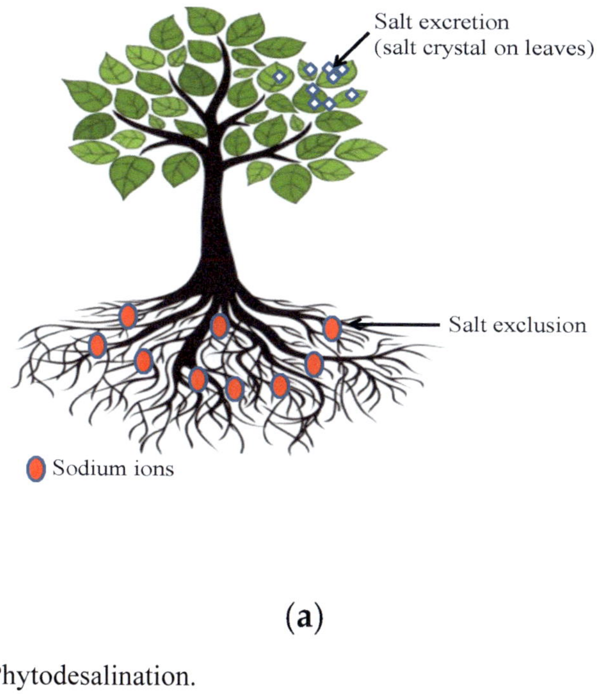

Figure 4.4. Phytodesalination.

Source: *Khan S, Masoodi TH, Pala NA, Murtaza S, Mugloo JA, Sofi PA, Zaman MU, Kumar R, Kumar A. Phytoremediation Prospects for Restoration of Contamination in the Natural Ecosystems. Water. 2023; 15(8):1498. https://doi.org/10.3390/w15081498*

4.7 HALOPHYTES FOR SALINE SOIL TREATMENT

The process of using halophytic plants to remove salts from saline soils so that the soils can support typical plant development is known as phytodesalination

(Figure 4.4). In comparison to glycophytic plants, halophytic plants may be more naturally suited to deal with heavy metals (Manousaki and Kalogerakis, 2011). They have systems in place, such as osmotic adjustment and ion compartmentalization in their vacuoles, that allow them to adapt to life in a salty environment, such as seawater, a saltwater marsh, or a salt desert. There is no accepted definition for halophytes, even though their great tolerance to salinity is acknowledged as a common trait of these species. The ability of plants to thrive in soil that is hazardous to other plants is known as heavy metal tolerance in plants. This is a result of a genetic interaction between the environment and the genotype. Physical barriers act as the first line of defence for plants against metals. The function of the plant cell wall and its binding capabilities in connection to metal tolerance is poorly understood. A substantial buildup of metal between the cell wall and the cell membrane was documented. Plant cell walls include functional groups like -COOH, -OH, and -SH that enable divalent and trivalent metal cations to bind to plants. The pectin, which is made up of carboxyl groups, is the most crucial part of the plant cell wall. Divalent and trivalent heavy metal ions attach to the carboxyl group of pectin when exposed to metal stress.

4.8 ENDOCYTOSIS: THE IMPORTANCE IN REMEDIATION

Endocytosis entails attachment to JIM5-P inside the cell or remobilization along with the pectin epitope. Cell walls of *Minuartiaverna sp.* growing on mine dumps have been found to contain elevated levels of Fe, Cu, Zn, and Pb. As the first cell structure to be exposed to heavy metals, the plant plasma membrane serves as the line of defence against heavy metal contamination. Metal uptake and accumulation are limited by the plasma membrane because it prevents metals from entering the cytoplasm. By altering the ability of the cell wall to bind metal ions, the membrane ion channels, the efflux pumps, root exudates, modified ion channels, or modified efflux pumps, metals can be restricted to the plasma membrane (Chandra and Kumar, 2017b). One strategy to tolerate or reduce metal toxicity is through the biosynthesis of various cellular macromolecules. This entails the induction of a wide range of low-molecular-weight protein metallochaperones or chelators, such as nicotianamine, putrescine, spermine, mugineic acids, organic acids, glutathione, phytochelatins, metallothioneins, or cellular exudates, such as flavonoid and phenolic compounds, protons, heat shock proteins, specific amino acids, such as proline. The equilibrium of the cellular redox systems in plants is upset when the aforementioned methods to prevent metal poisoning are not accessible, which increases the induction of ROS. Plant cells have evolved an antioxidant defence system to counteract the damaging effects of free radicals. This system is made up of both enzyme-based antioxidants like superoxide dismutase, catalase, ascorbate peroxidase, guaiacol peroxidase, and glutathione reductase and non-enzymatic antioxidants like ascorbate, glutathione, carotenoids, alkaloids, tocopherols, and proline.

Heavy metals like Mg, Fe, Mn, Zn, Cu, Mo, and Ni, which are necessary for plant growth and development, can be accumulated by all plants. Some plants have the capacity to collect metals like Cd, Cr, Pb, Co, Ag, Se, and Hg that are not known to have any biological purpose in plants. Hyperaccumulators are types of plants that combine a very high level of tolerance to foliar accumulation of trace amounts of heavy metals and other elements. Due to their possible application in phytoremediation, hyperaccumulators have recently attracted a great deal of interest. A hyperaccumulator can build up exceptionally high levels of trace metals in its aerial parts without exhibiting harmful symptoms. To avoid the negative consequences of metal toxicity or deficiency, plants have evolved adaptable mechanisms to deal with changes in their environment. Membrane transporters are needed in absorption, efflux, translocation, and sequestration of both essential and non-essential mineral elements, plant responses might include a major modification in gene expression (Suresh and Ravishankar, 2004). Metal hyperaccumulator plants have a number of distinctive qualities, including the capacity to absorb and transfer extremely large amounts of metals to their shoots and tolerate the toxic metals. The Zn-regulated transporter, Fe-regulated transporter-like proteins, natural resistance-associated macrophage proteins, and the MTP family of proteins are protein families whose members can mediate the transit of metals into the vacuole or other subcellular compartments. Toxic ions are removed from sensitive locations through chelation and sequestration mechanisms. Most of the hyperaccumulated metals are thought to be attached to ligands such as proline, citrate, malate, glutathione, phytochelatins, metallothioneins, or nicotianamine, which are important in the detoxification and sequestration of metals. Organic pollutants may be removed by plants by phytoremediation directly by absorption, translocation into plant shoots, metabolism, volatilization, or phytovolatilization, or indirectly by plant-microbe-contaminant interactions in plant root zones. The important stage in the metabolism of organic molecules by plants is uptake. Since organic chemicals are typically xenobiotics, their uptake is typically accomplished by simple diffusion through cell walls and membranes. Unfortunately, there are no transporters for organic substances. Organic substances may become sorbed to the surface of roots when they come into contact with them. The potential absorption of organic molecules is also influenced by their hydrophilic and hydrophobic properties. Effective hydrophobic organic pollutant binding is possible with hemicellulose in plant cell walls and the lipid bilayer of plant membranes. Plant enzymes add functional groups, such as hydroxyl groups (-OH), to the xenobiotics after they have been absorbed, increasing their polarity. After being absorbed and translocated, an organic component passes through one or more phases of change. The xenobiotic can be transported easily through the aqueous channels because of the enhanced polarity.

The xenobiotic is sequestered within plant vacuoles or by attaching to the plant cell wall and lignin via ATP binding, ABC transporters, and multidrug resistance proteins during the final stage of phytotransformation, which lessens the toxicity of the foreign molecule (Hwang et al., 2016). The xenobiotic has the ability to polymerize in a way that is similar to lignin and create a complicated structure that is hidden inside the plant. Surface soil and water contamination in the environment are primarily caused by complex industrial waste that is produced by various processing industries, including distilleries, pulp and paper mills, textile factories, tanneries, and petroleum processing plants. This complex waste is distinguished by its high amounts of phenol, sulfate, heavy metals, total dissolved solids (TDS), biological oxygen demand (BOD), chemical oxygen demand (COD), and other complex organic contaminants. High concentrations of heavy metals and other complicated recalcitrant organic contaminants make it difficult to safely dispose of complex industrial waste in the environment. The most accurate approach for predicting the availability of a contaminated metal to plants and the application of phytoextraction methods to remove heavy metals from polluted locations is the degree to which native plants are able to take up an element. The phytoremediation potential of using such plants at contaminated sites as natural hyperaccumulators of heavy metals from complex organic wastes has been demonstrated by a number of prospective wetland plants that grow on distillery waste-contaminated sites under natural settings. From washing fibers to bleaching, dyeing, and cleaning final goods, the textile business consumes a lot of water in all of its processes. Textile dyes and dye intermediates with high aromaticity and low biodegradability have emerged as major environmental pollutants. Textile effluent and wastewater from dye-synthesizing processes are two examples of inadequately treated wastewater that include organic colors. About 10 to 15 percent of the dye used in the textile industry is lost in the effluent during the dying process (Sahu and Singh, 2019). The significant amounts of wastewater produced also include a variety of chemicals employed during processing. High levels of BOD, COD, TDS, TSS, and color in the effluents from the textile industry may alter the water's quality, cause an odor, and severely impede economic activity, making their effective treatment a top priority.

As they are toxic and persistent in the environment, heavy metal-based fertilizers, fungicides, and pesticides are of primary ecological concern because they may eventually bioaccumulate to higher quantities and endanger humans and other living things. Pesticide application has the potential to spread from the applied region to non-target areas far from the crop, having an impact on not just pest species but also perhaps nontarget endangered species. It has been demonstrated that a variety of variables influence the phytoremediation of pesticides. It has long been known that the capacity to take up and translocate heavy metals is not only species-specific but also ecotype-specific. The processing of petroleum products is expanding quickly, which has led to the production of massive volumes of waste that constitute a severe problem to environmental quality globally. Petroleum hydrocarbons (PHCs) are a diverse group of hydrocarbons that are commonly found at petrochemical sites and storage areas, waste disposal pits, refineries, and oil spill sites. These hydrocarbons include linear and branched alkanes, cycloalkanes, aromatic hydrocarbons, and more complex chemicals. PHC buildup today has a significant impact on both human health and environmental safety. As a result, the rehabilitation of PHC-contaminated soils has garnered attention on a global scale. Numerous plants have been chosen because of their potential to speed up the phytoremediation of petroleum hydrocarbon-contaminated environments. Because of their large, fibrous root systems, prairie grasses are considered to be the best transporters for phytoremediation. Using genetically modified plants designed specifically for phytoremediation is one method to address these limitations. Phytoremediation research is interdisciplinary and necessitates prior expertise in environmental engineering, ecology, soil chemistry, and soil microbiology. The interactions between the primary components of the rhizosphere, which include metals, soil, bacteria, and plant roots, should be further investigated. The main goals of waste management are to lessen the negative impacts of pollution on both human health, as well as to potentially recover valuable materials from the waste. To manage waste and preserve the environment, a number of conventional techniques are used, including composting, fluidized bed combustion, plasma gasification, incineration, and combustion. These techniques also waste a lot of energy, cost a lot of money, pollute groundwater or the air, and even wipe out microbial communities that break down waste. In order to achieve an environmentally friendly waste management system, it is important to decrease the current levels of output and enhance material and energy recovery.

4.9 PHYTOREMEDIATION DETAILS

Given that phytoremediation is based on the employment of green plants to collect, sequester, and detoxify contaminants, it is often regarded as an ecologically responsible alternative to the environmentally damaging physical clean-up methods currently in use. Based on the effectiveness of metal translocation to the plant shoots via the symplast and apoplast, sequestration of metals within cells and tissues and transformation of accumulated metals into metabolically less harmful forms, hyperaccumulator plants use various metabolic processes for the mobilization and uptake of metal ions from soils. The plant biomass is collected, dried, or burned at the end of the growing phase, and the contaminated material is dumped in a designated area. If the exhaust emissions from burning the biomass can be properly cleaned, the energy generated by the process can sustain. The term metallophytes, or plants that grow on metal-enriched soils, refers to a subgroup of the category known as hyperaccumulators. However, due to their poor biomass production and delayed growth, these species cannot be utilized in phytoremediation. In order to rehabilitate heavy metal-contaminated locations, high-biomass crops and/or trees are therefore being examined as alternatives to hyperaccumulator plants.

The ideal plants for the phytoextraction of heavy metals from any polluted site should possess the following traits. They should be tolerant to high metal concentrations. They should accumulate reasonably high metal concentrations in their above-ground tissues and they should have a rapid growth rate. They should produce reasonably high biomass in the field and possess a profuse root system. The degree of metal transfer from root to shoot varies among plant species. The majority of plant species that have been designated as hyperaccumulators so far meet the aforementioned requirements. Additionally, the bioconcentration factor (BCF) and the translocation factor (TF) can both be used to determine the potential for phytoextraction (Das, Goswami, and Talukdar, 2014). An ecological and physiological response to metalliferous soils is metal hyperaccumulation. In order to understand how and why plants developed the ability to accumulate metal, ecologists have been studying the ecological effects of metal hyperaccumulation. These studies have determined the adaptive value of hyperaccumulation and looked at how the extreme metal concentrations in hyperaccumulators affect the interactions between different species of plants in the environments where they have evolved. Any plant that needs to be removed to have a reduced impact on the environment, economy, human health, or amenity is considered a weed. Invasive plants are also weeds. They are frequently the first species to colonize and dominate in disturbed areas because they are so good at surviving and reproducing there. A plant that colonizes and persists in an ecosystem where it did not previously exist can be an exotic species. When factors within their natural environment alter and make it possible for them to more effectively compete with other species and/or expand their population size and/or density, native plants can also turn into weeds. Weed species exhibit more environmental tolerance as compared to other natural plants. They can withstand a variety of stresses, including severe metal tension. For phytoremediation, it is, therefore, preferable to identify the species of weed plants present at a location. Under unfavorable environmental conditions, weed species typically have larger biomass and are quick-growing plants. Most contaminated locations have tolerant weed species, and phytoremediation can prevent the contaminants from entering the food chain by using these weeds and other non-edible species.

A wetland is a piece of land that has been continually or occasionally inundated with water to the point where it resembles a certain habitat. The most significant ecosystems on Earth are wetlands. However, wetlands have the innate ability to trap and/or effectively change a wide range of pollutants, acting as natural filters. Wetland plants, also referred to as macrophytes or hydrophytic plants can grow in contaminated settings. These include aquatic mosses and vascular plants. The presence of metal-accumulating plant species, which can accumulate a wide range of metals and so serve as affordable alternatives for reducing pollution loads, is one such feature of wetlands. In a contaminated area, these plants can collect metals, halting their proliferation and preventing their transfer into food chains. Constructed wetland (CW) treatment systems include wetland plants as a crucial component. Compared to technological wastewater treatment procedures, CW systems provide an affordable option (Herath and Vithanage, 2015). Depending on their ability to break down and remove nutrients and other organic and inorganic pollutants from contaminated sites, wetlands plants perform key functions in a CW. Wetland plants can either eliminate or stabilize contaminants from wastewater and soils in CW and natural systems. Plants and microbes are present, and their actions contribute to a significant portion of CW treatment. Since CWS and floating plant systems have been widely utilized for many years to treat various wastes, this idea is not new. Vegetable crops should not be grown on contaminated soil for phytoremediation due to the high levels of heavy metals in these crops, which have caused environmental health risks through food chain pollution. Due to their persistence and accumulation from contaminated environmental areas, a number of heavy metals worsen human and animal health. These metals are known to have carcinogenic effects on both people and animals as well as to induce renal diseases, asthma, and brain toxicity.

For use in phytoremediation of heavy metal-contaminated locations, trees may be the most affordable, environmentally responsible plant species, especially when alternative treatments are not cost-effective. It will be less expensive to develop trees in areas with poor soil structure and low fertility. Compared to the majority of herbaceous plants, trees have root systems that reach several meters deep into the ground. Salix and Populus species also offer a lot of promise for use in the creation of phytoextraction techniques.

Due to their broad root systems, high rates of water intake and transpiration which result in effective movement of chemicals from roots to shoots, quick growth, and high biomass production, salix trees are fast-growing trees and suitable candidates for phytoremediation applications. Populus trees are adaptable to a variety of climates and are increasingly used in forest systems for the manufacture of pulp and paper. This opens up the prospect of employing transgenic poplar plantations throughout a number of multiyear cycles to remove heavy metals from contaminated soils.

A thick canopy of trees would stop wind-driven soil contamination and erosion. Despite not accumulating significant amounts of metals, non-hyperaccumulator plant species can be employed in phytoremediation since their biomass production is many orders of magnitude more than that of common hyperaccumulator plant species.

Because of the connection between metal accumulation in aerial parts and growth capacity, overall metal extraction may be higher in non-accumulator plant species than in hyperaccumulator plant species. The synthetic aminopolycarboxylic acids (APCAs) trans-1,2-cyclohexylene dinitrilotetraacetic acid (CDTA), ethylenediamine-N,N′-bis(o-hydroxyphenyl)acetic acid, ethylenediamine-N,N′-bis(2-hydroxyethyl) iminodiacetic acid, and hydroxyl-EDTA have all been investigated. Because of its excellent efficacy in removing various metals, EDTA has been the most widely utilized chelating agent among all those employed for phytoextraction (Chandra, Kumar, & Singh, 2017).

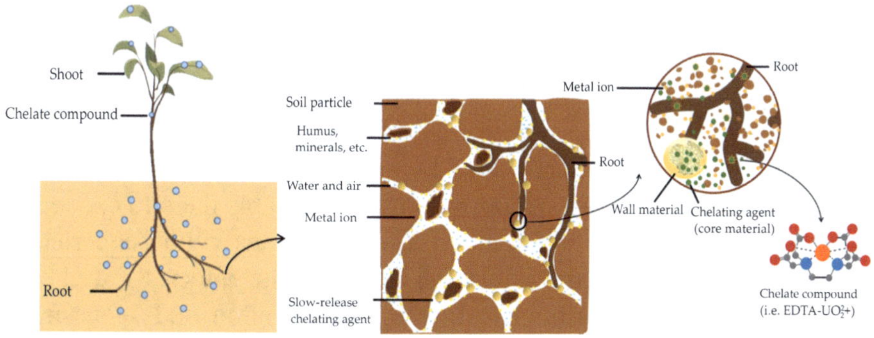

Figure 4.5. EDTA-facilitated phytoremediation.

Source: *You Y, Dou J, Xue Y, Jin N, Yang K. Chelating Agents in Assisting Phytoremediation of Uranium-Contaminated Soils: A Review. Sustainability. 2022; 14(10):6379. https://doi.org/10.3390/su14106379*

The amount of heavy metals absorbed by plants during EDTA-facilitated phytoremediation (Figure 4.5) is negligible in comparison to the amount mobilized from the soil and the significant amounts that leach out of the root zone. In EDTA-treated soil columns where cabbage plants were growing, significant amounts of Pb, Zn, and Cd had leached into the drainage water. Additionally, because EDTA and EDTA-heavy metal complexes have a poor level of biodegradability, they can stay in the environment and be hazardous to plants and soil microorganisms. Due to the unregulated solubilization and movement of metals, the in-situ application of EDTA could potentially increase the danger of water pollution. Further investigation is still required in order to reduce the use of EDTA and the accompanying possible risk of solubilized metal migration into groundwater. EDTA-heavy metal complexes are harmful to soil microbes and plants since they significantly reduce shoot biomass. However, the use of such synthetic chelating agents introduces an environmental risk due to their high mobility in the soil, which can cause the contaminant to be transported to nearby uncontaminated areas. The potential for excessive heavy metal uptake by plants, which raises the possibility of enhanced plant toxicity, is a crucial factor to take into account when using synthetic chelates. It is well recognized that the organic chemicals that roots exude may both directly and indirectly affect how easily poisonous and necessary ions dissolve. Phenolic acids are produced by roots and are found naturally in plants. They include acidic functional groups that are crucial to the chelation of heavy metals, such as carboxyl and phenolic groups. Hyperaccumulators effectively transport metal ions from the roots to the shoots through the xylem, in contrast to non-accumulator plants that retain the majority of heavy metals absorbed from a contaminated site in their root cells, detoxifying the metals via chelation in the cytoplasm, or store them in vacuoles.

The important characteristics of hyperaccumulators include their high efficacy in detoxifying and sequestration, which enable them to concentrate significant levels of heavy metals in aboveground organs without experiencing any phytotoxic effects. Various tissues, including the epidermis, trichomes, and even the cuticle, participate in the detoxification and sequestration of heavy metals. The distribution of metals to the trichome and cell wall, followed by the chelation of metals in association with ligands and the sequestration of the metal-ligand complexes into vacuoles, constitutes a typical mechanism for heavy metal detoxification in plants. Organic acids like citrate and malate, which are small ligands, are important detoxification agents. Such ligands may play a crucial role in evading the survival of heavy metals as free ions in the cytoplasm and, more importantly, in facilitating their entrapment in vacuoles, where metal-organic acid chelates are most frequently found.

The acidic environment of the vacuole is favorable for the development of metal-organic acid complexes. Using the model plant Arabidopsis or naturally hyperaccumulating plants, many of the genes involved in metal absorption, translocation, and sequestration have been discovered. The small size, slow growth rates, and constrained growing habitats of these natural hyperaccumulators, however, restrict their phytoremediation abilities. As a result, significant removal of the heavy metals may be accomplished if the genes were transferred to plant species like poplar and willow, which have large biomass and broad root systems. When employing genetically modified plants, this is especially true. Risk assessment is made more difficult by their prolonged life cycles, necessitating the need for more focused research in this area. Crops used for phytoextraction would be cut down before they went to seed, lowering the risk that they would cross with other crops used for food or get into the food chain. Wildlife may be harmed by plants that have been genetically modified to hyperaccumulate toxic metals in their leaves, yet studies show that people avoid such plants since their leaves don't taste good.

4.10 DETOXIFICATION USING PLANTS

Detoxification procedures, complexation by organic chelators and compartmentalization of metals in the vacuoles are some examples of these tolerance measures. We can better understand the adaptive mechanisms of plants and learn more about phytoremediation by understanding metal accumulation and histological changes in plant tissues brought on by heavy metal stress. Numerous phytoremediation studies have documented alterations in the anatomical, physiological, and morphological traits of plants as a result of their adaptation to environmental stress by heavy metals. The physiological and metabolic changes that occur in plants in response to high levels of heavy metals might be seen as potentially adaptive plant responses to stress. High metal concentrations that harm plant roots and impede nutrient uptake are likely to be the cause of morphological and biochemical reactions, which impede normal plant growth. Microscopic examinations of plants treated with high amounts of Zn showed deposits in roots and stems, destruction of parenchyma cells, and a drop in starch content in leaves. Zn-treated plants have significant reductions in relative water content, dry weight, and plant height among other physiological and morphological alterations. Only the stems and roots underwent structural alterations when Cd was present in greater amounts. The effects of Zn and Cd buildup on the anatomical, morphological, and physiological traits of barley. Significant reductions in water content, fresh dry weight, and plant height were also linked to Zn accumulation. At the high concentration, cadmium merely altered the structure of roots (Nedjimi, 2021).

4.11 PLASTIDS: THE EFFECTS OF HEAVY METALS

The plastid regions exhibited the most significant changes, which were then followed by modifications to the plasma membrane and nuclear architecture in all cultivars. Chloroplasts shrank in size and frequently had membrane breaks. Thylakoids grew larger, shorter, and fractured in various places. Following this, the plastoglobuli, starch granules, and lipid bodies grew in size and quantity. Plasmolytic shrinkage increased, although vacuolar compartmentalization was intensified. At this highest concentration, the ultra morphological alterations were more significant than at lower levels of Cd. With rising Cd levels, there were also more noticeable alterations in the size of vacuoles, the number of lipid bodies, and the thickness of cell walls. Plant cropping in contaminated soil, sludge, or sediment repeatedly until the metal concentration is reduced to a tolerable level is known as phytoextraction. A successful phytoextraction step will produce enormous amounts of potentially dangerous plant biomass (Ruiz et al, 2011). The issue of disposing of polluted plants and crops following phytoextraction is crucial for reducing the threats to environmental health. After harvestable metal-enriched biomass has been removed from a location, contaminated plants may be disposed of via landfills, deep well injection, or incinerated. Composting, compacting, and thermal processing are examples of post-harvest procedures that can decrease the volume and/or weight of biomass before it is disposed of as hazardous waste. The composting process produces soluble organic compounds that increase the solubility of lead. Composting can greatly reduce the amount of collected biomass; however, lead-contaminated plant biomass would still need to be treated before being disposed of. Care should be made to collect and dispose of the leachate in the event of compaction.

High amounts of heavy metals complexed with chelators after induced phytoextraction or Ni, Cd, and/or Zn after continuous phytoextraction in soluble and bioavailable forms will be present in the leachate produced by squeezing contaminated plant biomass. Utilizing hazardous biomass produced by phytoextraction through pyrolysis, incineration, and gasification is a successful method. Under anaerobic conditions and at moderate temperatures, pyrolysis breaks down the material. A lead or zinc smelter may use coke dust rather than coke, and lead or zinc could then be recovered throughout the smelting procedure. By destroying organic materials, the process releases metals, primarily as oxides. The released metals are discharged into effluent gases. The metal-containing dust is efficiently captured by modern gas cleaning technologies. An incinerator facility can also be used to eliminate plant material. Desiccation can be used to reduce the amount of plant material that needs to be delivered to a smelter. Drying lessens the possibility of plant material leaching during harvest and transportation. This kind of remediation can be made into a successful business when biomass generation, phytoextraction, and commercial energy consumption are coupled. The leftover ash is then ready for use as an ore. Thermal energy is a cheap and renewable resource that can be produced using phytomining as a potential energy storage method. Anthropogenic interferences in the biosphere have resulted in dangerously high levels of resource contamination with subsequent hazardous effects on natural biochemical cycles. Examples of these interferences include the accelerated rates of industrialization and urbanization, changes in agricultural practices, and extensive mining activities. Typically, plants are the source of heavy metals entering the food chain. Plants absorb heavy metals through contaminated soil, irrigated water, or air deposition on specific plant parts. Whether they are necessary or not, heavy metals have a major impact on how plants develop. Plants require essential heavy metals, but only in very minute quantities. However, non-essential heavy metals may be hazardous to plants and interfere with their regular physiological and metabolic activities. By forming connections with the sulfhydryl groups of amino acids, heavy metals can either alter the structure of protein molecules or they can render crucial enzymes and cellular components dysfunctional. Heavy metals can damage plasma membrane architecture, which can negatively impact crucial plant functions like photosynthesis, respiration, and enzyme activity. Reactive oxygen species (ROS) are produced more frequently when heavy metals are present, which disrupts the cellular balance of plants and has detrimental effects.

Plants have evolved enzymatic and nonenzymatic antioxidants to deal with heavy metal-induced ROS generation in an effort to combat the harmful effects of heavy metals. The glutathione (GSH) family of enzymes are among the several antioxidants that are triggered by heavy metal stress and play a crucial role in both the direct quenching of ROS and the detoxification of methylglyoxal (MG), which is formed in excess as a result of heavy metal stress. Phytochelatins, metallothioneins, proline, and other specific biomolecules are synthesized by plants as part of a few physiological tactics to counteract heavy metal stress in addition to their biochemical reaction. These biomolecules boost plant resistance to the effects of heavy metal stress. In some circumstances, plants also develop morphological adaptations that lessen the uptake of heavy metals from the soil, such as connection with mycorrhizal fungus. Plants use a variety of phytoremediation techniques in addition to these adaptations to cope with heavy metal stressors. Both natural and manmade sources release heavy metals into the environment. Various anthropogenic activities, such as improper irrigation and agricultural practises, unmanaged disposal of industrial effluents, the mining and metallurgical industries, the transportation of organic waste, the combustion of fuels, and the production of electricity, are generally responsible for the contamination of the environment with heavy metals. The most significant contributors to heavy metal pollution in the environment are industrialization and urbanization. Large amounts of heavy metals are also being introduced into aquatic and terrestrial ecosystems as a result of improper municipal solid waste management.

4.12 TOXIC EFFECTS OF HEAVY METALS

Toxic quantities of heavy metals interfere with normal physiological and metabolic processes. Various heavy metals, however, can have different sites of action within a plant. Chlorosis, (Figure 4.6) necrosis, impaired photosynthetic machinery, and decreased plant development are some of the most typical symptoms of heavy metal toxicity. Heavy metals can cause ultrastructural, biochemical, and molecular alterations in plant tissues that might accelerate the senescence process and ultimately cause plant mortality. Because of their capacity to attach to the sulfhydryl group of cysteine residues, heavy metals have the ability to inactivate enzymes.

Figure 4.6 Chlorosis: Morpho-physiological effects of Fe deficiency in efficient (EF) and inefficient (INF) soybean lines. (**A**) Total dry weight (DW); (**B**) chlorosis symptoms; (**C**) Fe concentration (µg/g) in roots and trifoliate leaves

Source: *Santos CS, Ozgur R, Uzilday B, Turkan I, Roriz M, Rangel AOSS, Carvalho SMP, Vasconcelos MW. Understanding the Role of the Antioxidant System and the Tetrapyrrole Cycle in Iron Deficiency Chlorosis. Plants. 2019; 8(9):348. https://doi.org/10.3390/plants8090348*

For instance, Cd binding to sulfhydryl groups on structural proteins can cause misfolding and limit enzyme activity as a result of Cd toxicity (DalCorso et al, 2008). The displacement of necessary metal ions from certain binding sites might interfere with crucial biological processes. Based on their mode of action, heavy metals can be separated into two groups. Heavy metal poisoning is thought to primarily affect plasma membranes. The thiol group is good for protecting plants from the oxidative stress brought on by heavy metals due to its reactive nature, relative stability, and high solubility in water.

CHAPTER 5

Management of Contaminated Soils

CONTENTS

5.1 INTRODUCTION

The process of cleaning up contaminated soils, sediments, and water is known as phytoremediation. Phytoremediation is a technology that aims to handle a variety of surface pollutants, in contrast to many conventional remedial engineering procedures. Targets for phytoremediation include industrial waste products, pesticides, explosives, petroleum hydrocarbons, polluting metals, and metalloids. The large cost savings that these systems seem to provide is the main factor driving further development in this field. However, because plants are living things with precise oxygen, water, nutrient, and pH constraints that must be maintained, phytoremediation has intrinsic limitations. Despite these drawbacks, a variety of phytoremediation techniques have emerged from the lab and are already in use. Commercial phytoremediation systems are currently in use to clean up waterborne pollutants and shallow aquifers. Advanced trials as well as field tests for the phytoextraction of metals from polluted soils are in progress. The majority of the present procedures are technically sound but far from being at their best. Although field testing has typically received favorable regulatory and public acceptance, many of them still have room for development and extension. Contaminant loading of water and air has typically decreased as a result of these changes. Contaminated soils may not exhibit significant inherent improvement over many decades due to poor kinetics and absent or decreased dissipation processes. Unfortunately, at many of the larger sites that require remediation, polluted soil is frequently the main issue. If unchecked, off-site migration of the toxins at such sites may continue to have an impact on groundwater, nearby areas, and bodies of surface water. Renewing these sites is proving to be time-consuming, and expensive. Remediation can mean one of two things: clean-up, in which the contaminant is either removed from the matrix through leaching, bioremediation, etc. or the entire contaminated matrix is removed from the site through excavation and landfilling at another site or stabilization, in which the matrix or contaminants physical or chemical form is changed to one that is more inert. The desired treatment must be legally compliant, technically and logistically practicable, completed in the allotted amount of time, and economically viable (Vangronsveld et al., 2009).

5.2 TRADITIONAL TECHNIQUES

Traditional techniques for cleaning up contaminated groundwater, sediments, and soils frequently rely on civil and chemical engineering innovations from the past decades. Many different physical, thermal, and chemical processes fall under this category. However, there are some situations where biological processes particularly microbial have proven useful. The legal standards for environmental clean-up have recently been more lenient, which has boosted the acceptability of such passive remediation methods. However, remediation strategies created do not rely on natural attenuation. Engineering technologies can operate in a wide range of oxygen, pH, pressure, temperature, and osmotic potentials and are frequently faster and less sensitive to heterogeneity in the contaminant matrix. The cost of employing efficient engineering techniques is frequently unaffordable, which is viewed as a benefit of bioremediation.

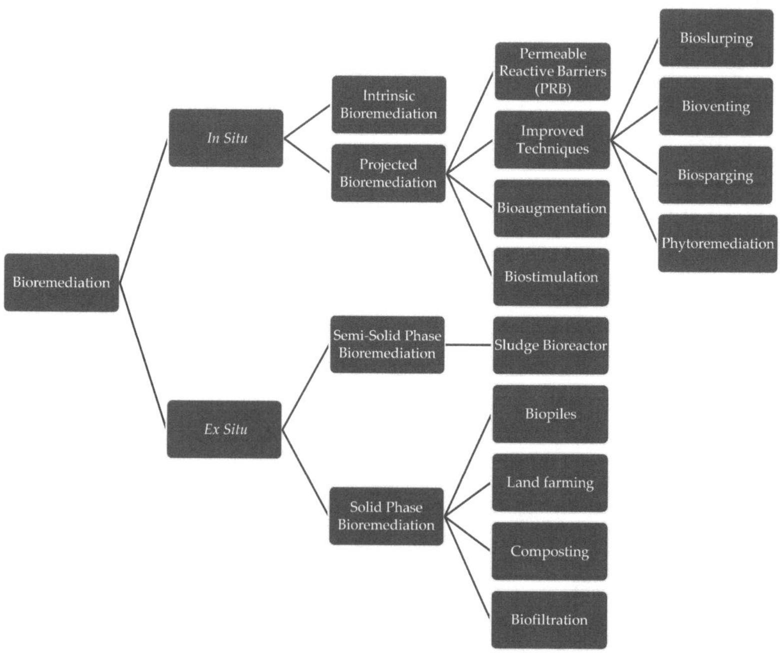

Figure 5.1. Soil remediation: overview of technologies.

Source: *Sales da Silva IG, Gomes de Almeida FC, Padilha da Rocha e Silva NM, Casazza AA, Converti A, Asfora Sarubbo L. Soil Bioremediation: Overview of Technologies and Trends. Energies. 2020; 13(18):4664. https://doi.org/10.3390/en13184664*

It would seem that alternative solutions based on biological activity would not be developed if clean-up using conventional technology could be done at a reasonable price. Numerous agricultural practices may be useful for soil remediation (Figure 5.1) because the principal objective (agriculture) is the preservation of soil quality. The majority of agricultural soil management techniques rely on the biological activity of the plants and the related microbial communities as well as the physical protection provided by a vegetative cover. Stabilization and biological activity form the foundation of plant-based methods for the removal of both organic and inorganic environmental pollutants.

5.3 AGRONOMIC METHODS

Agronomic methods for stabilizing polluted sites in situ are still being developed. To lessen the bioavailability of the pollutants, soil supplements are supplied to contaminated soil. After the site is vegetated, the stabilized soil matrix and contaminants are less likely to migrate off-site due to water use and erosion control. In order to sequester the material in the root tissue with minimal translocation to the shoot and potentially aid in the catalysis of the formation of insoluble contaminants like Pb-pyromorphite, plants are selected to maximize root uptake of any small amounts of contaminants that would escape from the stabilizing mix. Numerous instances support the viability of this approach. For instance, it is well recognized that Pb from natural sources such as soil and dust is frequently less physiologically accessible than Pb from anthropogenic sources with comparable quantities. Both soil solutions and mammalian gastrointestinal tracts have trouble digesting these natural forms. It has been demonstrated that Pb is capable of being converted to less dangerous forms in a natural setting. More crucially, compared to most engineering solutions, these methods are less likely to impair the soil's capacity to support plant development. In order to restore damaged regions, vegetation has long been employed, and tolerant vegetation is frequently seen on or planted into contaminated soils. It makes sense to look into using plants to clean up contaminated soils, aquifers, and wetlands given extensive experience with planted soils in agriculture and the more recent problems of environmental contamination. Soil remediation is a complicated technical procedure (Shen et al., 2022). Equivalent pollutant concentrations in soils frequently have toxicity effects that differ greatly. The availability of the pollutant and its potential for damage is influenced by a number of factors, including contaminant speciation, soil pH, other ions in solution, kinds of clays and oxide surfaces, the presence of organic matter, vegetation, and rainfall. Remedial action is often required when anthropological activities pose an unacceptable danger.

It is challenging to pinpoint the specific area of soils that require clean-up because risk assessment is an inexact science, and because community-level risk tolerance and economic resource availability vary greatly. Despite this ambiguity, even a quick glance at some sites reveals the need for clean-up. Sites that lack vegetation are likely to be a source of the passage of toxins and are likely candidates for remediation of some kind.

Normal soils have an elemental composition that depends on the geological and physical processes that took place during their generation. The soils produced from sea sediments are distinct from those produced from heavy metal-rich rock outcrops. Along with this natural variety, anthropogenic actions have made soil heterogeneous. The handling and use of fossil fuels, industrial manufacture, the application of fertilizers, and municipal sludge to land are the most frequent sources of anthropogenic inorganic contamination. The most challenging in terms of technicality are those involving heavy metal decontamination in sediments and soil. Numerous inorganic pollutants form a strong bond with the soil matrix. This is almost always true at older contaminated sites because more mobile pollutants have already left the site in regions with even light rainfall.

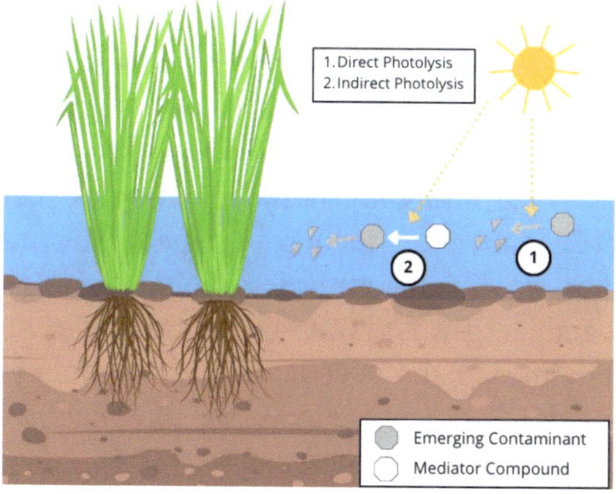

Figure 5.2. Utilization of wetlands for the photodegradation of emerging contaminants.

Source: Overton OC, Olson LH, Majumder SD, Shwiyyat H, Foltz ME, Nairn RW. Wetland Removal Mechanisms for Emerging Contaminants. Land. 2023; 12(2):472. https://doi.org/10.3390/land12020472

For the remaining contaminants, there are few clean-up alternatives available. However, the majority of site clean-ups entail excavation and transportation to a safe landfill region. Some advanced treatments include soil washing or electroosmosis. Alternately, some of these locations have also been capped, or simply covered. Utilizing plants to remove toxins from the environment is known as phytoextraction. The phrase was first almost solely used to describe heavy metals in soils, but it has now expanded to include many other substances.

5.4 WETLANDS

The early utilization of wetlands (Figure 5.2) for water purification was the catalyst for the studies on phytoextraction of inorganic contaminants in the environment. These systems made it challenging to recover the inorganic contaminants since they frequently precipitated out of the water and into the sediments. Unfortunately, especially in temperate zones, these systems are neither very effective nor affordable. More recently, field tests for radionuclides and heavy metals have been conducted on hydroponic designs based in greenhouses employing terrestrial plants. These systems were created using plants chosen for having low translocation to the shoots and high pollutant root uptake/affinity. These terrestrial plant-based applications were created to replace synthetic resin chelates for the filtration of water. Harvesting root biomass allows the pollutants to be eliminated from the system. Many biological restrictions built into phytoextraction, such as the poor transfer of many pollutants from root to shoot, are effectively accommodated by rhizofiltration. These technologies seem promising and demand less R&D innovation than soil remediation methods (Agostini et al, 2003). Other wastewater purifying technologies are currently very competitive. For reducing waterborne contaminants to the levels required by regulatory guidelines, many excellent non-plant solutions are available. The motivation to embrace new technologies may be low for facilities that are already compliant with local laws.

5.5 SOIL COMPONENTS VS. HEAVY METALS

The main components of soil are Si, O, and Fe-based minerals, which are challenging to distinguish from polluting inorganics. Phytoextraction is based on the idea that plants can be utilized to remove inorganic pollutants from the soil matrix and transfer them into a matrix made mostly of carbon, the plant material. Engineering methods are more easily applied to plant material when compared to those that were ineffective on the soil matrix. Many of these inorganics are extracted from the plant matrix using chemical, thermal, microbiological, and leaching processes. All plants accumulate a wide range of mineral elements, but they are generally quite good at rejecting those that are not necessary for their development and survival. The first exception is that some non-essential and comparatively non-toxic elements are absorbed in substantial quantities by specific kinds of plants. The element Si, which is absorbed by many plants to amounts that may reach 1 percent of their dry weight, is the best example of this. The second exception is that all plants, to some extent, reflect their environment of growth. This fact has been acknowledged by many geologists, who have employed plant-tissue analyses to find hidden mineral deposits. This method of using plants to find ore deposits containing elements like uranium is known as geobotanical prospecting. A limited subset of plants called hyperaccumulators have the capacity to absorb, translocate, and endure concentrations of specific heavy metals that are hazardous.

With very few exceptions, practically all plants exist within a small range of relative elemental concentrations in soil, althoughsoil concentrations of the majority of elements vary significantly. On the other hand, hyperaccumulating plants can absorb, translocate, and tolerate heavy metal concentrations in shoots that are higher than 0.1 percent Ni, Co, Cu, Cr, Pb, or 1 percent Zn on a dry weight basis. These plants, which frequently originated on metalliferous outcrops, are prominent for their high levels of accumulation and tolerance as well as for their almost avid urge to concentrate these metals from even normal soil. Despite the fact that many laboratories continue to concentrate on the research of these plants, effective phytoremediation with these species has remained elusive. Although the majority of locations exhibit mixed metal contamination, many of these species appear to hyperaccumulate just one metal. Many of these species grow slowly and produce little biomass while having high metal concentrations. This is particularly problematic because the amount of contamination removed per acre per year is frequently used to gauge the effectiveness of phytoextraction. No matter how fascinating biologically, slow-growing, low-biomass hyperaccumulating plants are, they cannot serve as the foundation for a technology that can be used for biomining or remediation. But hyperaccumulators continue to be a major source of inspiration for the molecular biology and agronomy communities. These plants demonstrate that biological systems can be created while keeping up to 4% metal content in tissues without suffering appreciable yield losses.

It is well known that many germplasm sources exhibit significant variation, and testing these for metal absorption, translocation, and tolerance has not been an exception. However, in parallel with these initiatives, researchers are looking into the underlying phytoextraction limitations. The most prevalent heavy metal pollutant in the environment, Pb, has been used in a significant portion of research. Three restrictions were found in this situation. The first was that, despite high total Pb concentrations, Pb soil solution levels were low. The second was that while most plant roots absorbed Pb efficiently, there was a poor transfer from roots to shoots. Lastly, biomass was significantly lowered by tolerance to Pb in plant tissue at high rates. Chelating solutions have been employed for a while. They have also been utilized in agronomic and horticultural situations to supply micronutrients to plants. They raise the solution concentration of numerous heavy metals in soils Many scientists were concerned about harming the plants by using excessive chelates or chelate-Pb complexes. Plants are cultivated in soil, chelates and other supplements are supplied, and Pb complexes are absorbed into the plant under the plant transpiration gradient. The Pb-chelate complex proceeds up into the plant shoots through the xylem. Since a plant spends most of its life without significant tissue Pb, plant tolerance is less relevant in this situation. After the soil is amended, the plants can be harvested in one to two weeks. Despite considerable harm to the plant, it is still possible to easily harvest the Pb-containing plant regardless of its physiological condition. The Pb phytoextraction industry currently offers the economic prospect. This pollutant has some severe restrictions due to its chemical and biological composition (Chandra and Kumar, 2017a). From a biological and chemical standpoint, targets like zinc or nickel would have been preferred. It is obvious that some technique of enhancing Pb solubility near the root zone will be necessary. Internal chelating mechanisms will be required to allow Pb to pass from root to shoot tissue and avoid being sequestered in the root tissue. The development of phytoextraction as a mining or remediation process has only involved a small number of plants. The biological potential is still mostly untapped since no plant has yet been chosen for maximal metal output in recorded history.

The general course of xenobiotics in plants is well understood. Plants generally have a wide range of metabolic capabilities that can influence what happens to a chemical after it enters the plant. In addition to pesticides, PCBs and, more recently, TCE have been used extensively in plant metabolism. A large portion of the contaminants that enter plants are integrated into the cell biomass, which is difficult to remove and identify chemically. The rate and size of these metabolic pathways in relation to the plant's overall pollutant uptake rate are still poorly understood. The rate of transpiration, the mobility and volatility of contaminants may now be key factors in deciding the fate or partitioning of contaminants. There is mounting evidence that suggests the methods used to conduct these studies may have a significant impact on how well we understand how pollutants behave. Despite the fact that plants can frequently metabolize or sequester environmental toxins, they still have two major disadvantages. Because they are predominantly autotrophic, plants obtain their energy from the building of their cells from CO_2, light, water, and minerals. Plants metabolize a more limited range of chemical structures as a result than their microbial counterparts. Plants frequently transform xenobiotics into more water-soluble forms within the plant tissue. Examples of this transformation include hydroxylation and glycosylation. The parent substance or its chemically changed form is frequently then trapped in the cytoplasm or cell wall matrix. In contrast, the product of microbial metabolism is frequently converted to carbon dioxide, water, and cellular biomass. In many aspects, it makes technological, financial, and regulatory sense to combine microbial degradative processes with the structural functions of plants.

5.6 MICROBIAL COMMUNITIES IN THE RHIZOSPHERE

The plant may only serve as a support system for microbial communities and degradative microorganisms in soils contaminated with lipophilic organic substances. Although microbial activity in the rhizosphere has typically been linked to an increase in degradation, a mechanistic understanding of the process is still lacking. Understanding how plant roots affect the microbes that cause chemical transformations is necessary for the successful application of plant-microbial systems for the removal of a variety of pollutants. The complexity of the system is substantially increased by the addition of a microbial community component that varies in time and place throughout the length of a root and in the soil. By supplying substrates for microbial growth, allowing the assemblage of distinct communities on root surfaces, and by changing soil chemical and physical conditions like redox, pH, and the availability of inorganic nutrients, plants may be able to influence the microorganisms that break down organic contaminants. Pollutants that are absorbed by the roots of a plant may be changed by the plant and then redeposited into the rhizosphere in a different chemical form. Significant amounts of carbon are released into the rhizosphere by growing plants as amino acids, organic acids, sugars, and a variety of other structurally varied compounds. These molecules can make up anywhere between 15 and 40% of the carbon absorbed by photosynthesis (Oloumi, Mousavi, and Nejad, 2018). A variety of microbial communities are supported by the majority of these compounds, which are released in the zone of elongation right behind the root tip. In older root zones, where carbon becomes limited, nematodes and protozoa that feed on bacteria associated with the roots are present in the rhizosphere population. The result of this bacterial growth is the selection of microbes that can coexist in an oligotrophic environment. As new carbon substrates become accessible due to soil organisms feeding on the roots or at locations where lateral roots pierce the cortical tissue, certain subsets of the oligotrophic community go through further succession.

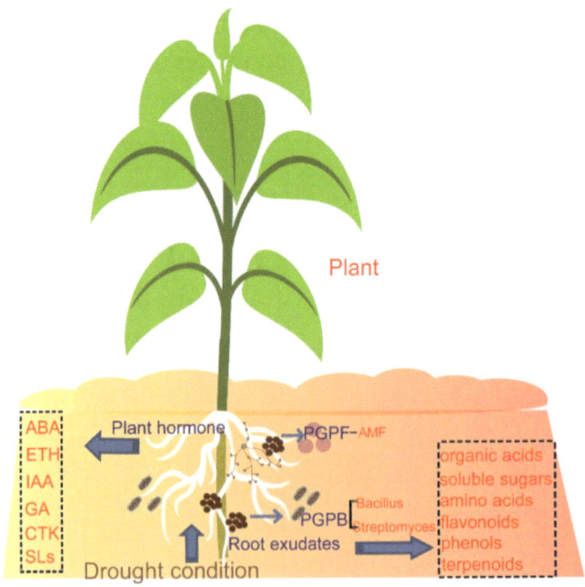

Figure 5.3 Drought stress, root exudates and associated microbiotas. ABA, abscisic acid; ETH, ethylene; IAA, auxin; GA, gibberellin; CTK, cytokinin; SLs, strigolactones; PGPB, plant growth-promoting bacteria; and PGPF, plant growth-promoting fungi.

Source: *Chen Y, Yao Z, Sun Y, Wang E, Tian C, Sun Y, Liu J, Sun C, Tian L. Current Studies of the Effects of Drought Stress on Root Exudates and Rhizosphere Microbiomes of Crop Plant Species. International Journal of Molecular Sciences. 2022; 23(4):2374. https://doi.org/10.3390/ijms23042374*

The composition of root exudates (Figure 5.3) that reach the rhizosphere microbiota is further altered by the establishment of mycorrhizae with symbiotic root-colonizing fungi in the older root zones. All of the chemicals produced by plants and their associated microbiota are eventually released into the soil as plant roots decay and die. They are destroyed by a different microbial community that includes specialized organisms that can feed on cellulose, chitin, and lignin. By bringing together dense populations of various microbes that have catabolic plasmids or genetic material, the plant may also encourage the emergence of degradative activity in the community. By serving as a mating surface for bacteria, plants may hasten the emergence of new catabolic pathways for the breakdown of xenobiotics because bacterial reproduction and genetic recombination depend on the active growth of microorganisms. Other plasmids may also transmit genes for chemo-attractant sensors, antibiotic or heavy metal tolerance, or other genes that are advantageous to the growth and survival of introduced organisms, in addition to plasmids containing catabolic genes.

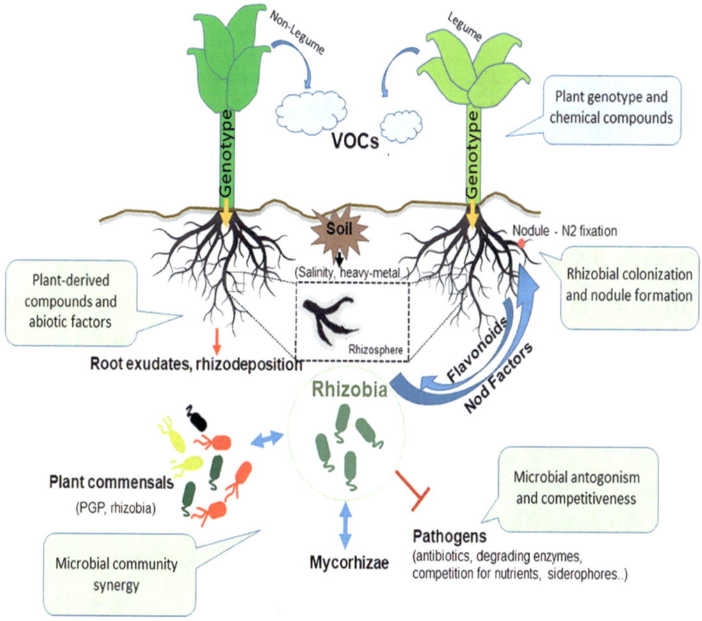

Figure 5.4. Rhizosphere and its interactions.

Source: *https://upload.wikimedia.org/wikipedia/commons/thumb/a/a2/Actors_and_interactions_in_the_rhizosphere.jpg/440px-Actors_and_interactions_in_the_rhizosphere.jpg*

It is well known that genes for catabolism of a variety of substrates are carried on plasmids, and recent research has shown that plasmid stability for degradation of a typically resistant substrate, 2,5 dichlorobenzoate, is increased in the rhizosphere (Biswal, 2022).

Since electrons are produced during the metabolism of root exudates, this could accelerate the pace of reductive dehalogenation. In addition to eliminating nitrate, which might replace oxygen as an alternative microbial electron acceptor, plants can also reduce the amount of oxygen in the soil. It may be possible to create conditions that are favorable to both aerobic and anaerobic microorganisms that could transform chlorinated compounds through sequential anaerobic and aerobic metabolism, depending on how water is managed in the soil or the presence of plants with aerenchyma to transport oxygen into the rhizosphere (Figure 5.4).

The microbial group would resemble a biofilm in terms of how the pollutant is distributed during bulk water flow. Understanding how the plant affects the microbial community is extremely difficult due to the dynamic nature of the rhizosphere.

Plant-microbial systems for bioremediation, may be especially useful for in-situ remediation of contaminated soils. Understanding the effects of plant roots on soil microorganisms as well as the advantages and drawbacks of using plants to boost the activity of either native or introduced xenobiotic degraders will be necessary for the successful application of plant-microbial systems against a wide range of contaminants that occur in the field.

The large population densities of various heterotrophic microbial species in the rhizosphere may facilitate the transformation of xenobiotics or provide an environment that is favorable for gene exchange and rearrangements. In some cases, structural analogues of different xenobiotics found in root exudates, cell wall components, and lysates, together with secondary products may promote the growth of microorganisms that metabolize xenobiotics. By boosting the mass flow of water-soluble compounds to the root surface, where the rhizosphere microflora can interact with them, evapotranspiration of water by plants may also have an impact on the transport of water-soluble compounds. If a microbial population is absent from the soil or is non-competitive in the rhizosphere, or if the degrader organisms can grow independently on the substrate, the presence of the rhizosphere may be unnecessary in other cases, the rhizosphere may have little or no impact on the degradation of xenobiotics. Knowing a molecule's chemical properties, as well as its pathway of degradation, and being aware of the impact of plant root growth on the microorganisms engaged in the degradation process, are necessary for predicting the rhizosphere effect for a given compound.

5.7 RHIZOSPHERE POPULATION

The variability of the rhizosphere population, which represents microbial succession following the initial colonization of new roots, can also be influenced by different root zones. Early research on the rhizosphere's microbial population revealed that pseudomonads, which make up anywhere between 30 and 90% of the culturable organisms isolated on agar media, were the most common bacteria associated with plant roots. Recent research demonstrates that the community of microorganisms associated with roots is diverse and that these communities can differ for the same plant species in different soils or even for distinct plant species when utilizing media for oligotrophic microorganisms. Depending on random processes that result in the successful colonization of a certain root by various microbes, microbial communities associated with plants may even vary from one root to the next. The structure and change of microbial communities in a biofilm that spreads radially from the root surface, redox and gas diffusion gradients, and other factors help to define the environmental conditions that influence microbial transformations of xenobiotics. The feeding of bacteria on the root surface by protozoa and nematodes causes the growth of microorganisms, the release of nitrogen mineralized from the microbial biomass, and changes over time in the amounts of substrates that are available to competing microorganisms in the rhizosphere.

One of the several mechanisms for the breakdown of nitro xenobiotics is the reduction of nitroaromatic compounds to their corresponding amino derivatives (Del Buono et al, 2020). Research on the nitrodiphenyl ether herbicide acifluorfen revealed that it rapidly converted to aminoacifluorfen before being incorporated into inaccessible soil components in suspensions of both soil and the rhizosphere. Due to microbial populations in rhizospheres, particularly Gram-negative bacteria, aminoacifluorfen was produced more quickly in soil. In soils managed by phytoremediation, nitroreduction by rhizobacteria is a crucial catabolic pathway for the breakdown of different nitroherbicides and other nitroaromatic chemicals. Because they fit a class of chemicals that are frequently used as herbicides, explosives, pharmaceuticals, and industrial chemicals, nitroaromatic compounds are a source of environmental concern. To reduce their persistence and eventual environmental pollution, better management strategies can be developed by better understanding their biotransformations. An environmentally preferable method of cleaning up contaminated matrices is phytoremediation. When using phytoremediation techniques, xenobiotics can degrade more quickly because of the uptake, transformation, and general rhizosphere influence, which strengthens the microbial community that can biodegrade the contaminant. Bacterial nitroaromatic chemical transformations may occur by oxidative pathways, reductive pathways, or partly reductive pathways. Prior to ring cleavage and full mineralization, nitrophenols can experience an oxidative nitrite release and hydroxyl group replacement. The nitroreductase enzyme mediates reductive processes by converting the nitro moiety to the appropriate amino derivative.

Acifluorfen is a common herbicide used in soybean, peanut, and rice farming to eradicate a variety of grass weeds. The amino derivative breakdown metabolite is found in soil, particularly in low-redox environments. There are few investigations on acifluorfen breakdown in the soil. In soil, its half-life can be between 23 and 122 days. Aminoacifluorfen is the main accumulating metabolite in both anaerobic and aerobic settings. Trifluralin, a nitroaromatic herbicide, has been shown to degrade more quickly in the rhizospheres of tolerant plants that have colonized pesticide-contaminated areas. The fate of other nitroaromatic herbicides in the rhizosphere is poorly understood. The presence of combinations of pollutants in high quantities may hinder biological techniques. Herbicides are the main contaminants at many locations, making it difficult to clean up these soils in situ using a phytoremediation technique. However, studies of soils from several sites that were contaminated with pesticides have produced positive findings, showing that some plants can withstand only mild contamination. It is critical to comprehend how the presence of other pesticides can affect how certain pesticides behave in soil. The limited bioavailability of the chemicals for subsequent biological system degradation is another issue with soils that have been poisoned over a long period. It is crucial to take into account the impact of soil aging while conducting laboratory experiments to understand the fate of pesticides in contamination scenarios. Enhanced deterioration is a phenomenon that can take place in soil that has been exposed to chemicals over an extended period of time (Haritash and Kaushik, 2009). Accelerated decomposition may take place when the chemical is applied to soil that has already been exposed to it. Although the preceding treatment of the soil causes this accelerated degradation, it does not occur in all soils or with all pesticides. When the substance is applied again, microorganisms that can endure or benefit from the selective pressure will predominate and be ready to destroy it more quickly. This is not ideal in a pest-management setting because failing to control invasive plants or insects might cost more money. However, in a case involving contaminated soil, this could be a beneficial trait because it would allow the chosen microbial population to quickly decompose successive spills. Soils that can quickly degrade a pesticide are the best places to look for populations of microorganisms that can be grown and used in bioremediation techniques like bioreactors or in situ inoculations. By absorbing toxins into the plant tissue and potentially converting them to harmless chemicals, plants can positively impact the elimination of organic waste. Because of the increased microbial activity linked to the roots, plants also affect how pollutants degrade. Soil from the rhizosphere of a herbicide-tolerant plant degraded a mixture of herbicides more quickly than soil from an unvegetated region.

5.8 SALINITY AND BIOSTIMULATION

The elevated salt content of the soil may have been caused by fertilizer leaks. Although transfer has been performed with carbamate and several organophosphate insecticides, the success rate of boosting biodegradation in soils by inoculation with known degraders has generally not been good. Environmental factors for intrinsic degradation are optimized through biostimulation. Under laboratory and field conditions, soil additives with the potential to biostimulate the breakdown of pesticides, particularly herbicides, have been investigated.

The ability of corn meal, sewage sludge, and maize and soybean residues to lower high levels of the herbicides alachlor, atrazine, metolachlor, and trifluralin was demonstrated using a combination of land farming and biostimulation.

Soybean meal or chicken litter was efficient at reducing high levels of trifluralin in soil under saturated moisture conditions, while only soybean meal was beneficial under field capacity conditions. Corn meal, poultry litter, or ryegrass residues were added to promote the decomposition of cyanazine and fluometuron, with ryegrass having the largest impact. When the herbicides were used at agronomic doses on soil, residue from winter ryegrass and soybeans, respectively, encouraged the degradation of metribuzin and fluometuron. Alfalfa residues delayed the degradation of metribuzin whereas glucose and wheat straw accelerated it.

The bioremediation of other pesticides, such as fungicides and insecticides, can also be increased by soil amendments. Supplemented with rice straw reduced the persistence of DDT in soil under anaerobic conditions. The addition of manures has encouraged the degradation of insecticides such as aldrin, dieldrin, and toxaphene.

5.9 PESTICIDES

Pesticides can be unusually persistent when present in the soil at high concentrations, possibly due to microbial toxicity that prevents biodegradation. Residue aging may potentially contribute to the persistent nature of pesticide residues. Herbicide absorption can increase over extended soil residence and may result in slower breakdown rates. It may be because older residues are less accessible than newly applied residues. To prevent local well water and neighboring properties from being contaminated by pesticide residues through leaching and runoff, pesticide-contaminated soil must be cleaned up. The two most popular methods of disposing of contaminated soil like landfilling or incineration can be costly and impractical for small-scale users like farmers. For *ex-situ* or *in situ* clean-up activities, bioremediation of pollutants has been suggested as a less expensive option. Composting and phytoremediation are two bioremediation methods that have been suggested for cleaning up pesticide-contaminated soils, while land farming has been employed in pilot trials and for cleaning up agrichemical locations. Bioremediation relies on boosting microbial activity that may be directly or indirectly linked to pesticide metabolism. Currently, several states have laws that make it legal to cultivate in pesticide-contaminated soils. The goal of land farming is to place contaminated soil in the uncontaminated A-horizon of soils where dilution would reduce the waste concentration just enough to promote both chemical and aerobic microbial breakdown. In order to promote microbial activity and effectively accomplish a bioremediation impact, standard agronomic management practices like tilling, fertilizing, and irrigation may be adequate. However, mixing contaminated soil with uncontaminated soil only may promote biodegradation, possibly in conjunction with an increase in the activity of soil dehydrogenase (Wang et al., 2008).

5.10 LAND FARMING

Land farming may be possible if the contamination is recent and the effective rates of application are very low. Land farming could be used to swiftly excavate and apply very thin layers of soil over uncontaminated land in the event of a minor pesticide leak. More acreage would be required for the treatment of highly polluted regions, especially if final pollutant loads had to be kept below typical field application rates. Since the land would typically be used for agriculture or other commercial activities, the speed at which landfarming could reduce pollutant residues to acceptable levels would be crucial in this situation. Pesticide residues in contaminated soil may have been present for many years previous to excavation or in-situ treatment, which could slow down the rate at which they would break down. We can combine soil dilution, which has previously been demonstrated to encourage the degradation of herbicides, with barley or soybean plant residue or corn meal additives to promote the cleaning up of herbicide-contaminated soil. Even without initially diluting the old soil, corn meal appeared to be a potent biostimulant for the breakdown of alachlor, metolachlor, and trifluralin. However, reducing the concentration of the pollutants in the soil would lessen the likelihood of translocation and phytotoxicity-related negative consequences. Under laboratory circumstances, atrazine breakdown could not be induced by organic additions, but it has been discovered that atrazine persistence is reduced when soil is biostimulated in the field. The disposal of soils polluted with pesticides might be feasible at a reasonable cost by combining organics with land cropping (Ward and Singh, 2004). Only if the waste pesticides were still legally registered for field application and not categorized as priority pollutants would this alternative be practical. To lower levels of some pesticide pollutants, phytoremediation is a desirable approach. Since they are inherently tolerant to toxins, several researchers have used plants that naturally colonize such contaminated regions.

Figure 5.5. Safeners.

Source: *Duhoux A, Pernin F, Desserre D and Délye C (2017) Herbicide Safeners Decrease Sensitivity to Herbicides Inhibiting Acetolactate-Synthase and Likely Activate Non-Target-Site-Based Resistance Pathways in the Major Grass Weed Lolium sp. (Rye-Grass). Front. Plant Sci. 8:1310. doi: 10.3389/fpls.2017.01310*

Safeners (Figure 5.5) are substances that shield crop plants from herbicide damage. They have a variety of chemistries and modes of action. To safeguard a variety of agricultural plants (usually monocots) against chloroacetamide phytotoxicity, many safeners have been created. Chloroacetamide herbicides can sometimes harm corn, especially when used in large doses early in the growing season and moist conditions. The enzyme glutathione S-transferase (GST) normally catalyzes the GSH-conjugation process to detoxify the chloroacetamide. Benoxacor increases GST in corn and is involved in the metolachlor detoxification process into a non-phytotoxic metabolite. In phytoremediation procedures, safeners like benoxacor may be useful to increase the tolerance of the remediation crop to high herbicide levels (Hoagland, Zablotowicz, and Locke, 1997). The primary metabolic pathway in the soil is dichlorination, which produces a number of acidic polar compounds. Fluorescent pseudomonads and other rhizosphere bacteria are involved in the detoxifying mechanism. Laboratory tests on soil treated with high concentrations of alachlor revealed that adding cornmeal and inoculating with *Pseudomonas fluorescens* strain UA5-40 might significantly speed up the initial breakdown of alachlor. The efficacy of using microorganisms that co-metabolize pollutants as bioaugmentation agents alone may be restricted. However, when used in conjunction with other bioremediation techniques, these microbes might be more useful.

A range of estuarine wetland plants have been cultivated, exposed to various herbicides under various circumstances, and then checked for signs of stress. The research was done to find out how these pesticides affected the native plant communities bordering these estuaries since pesticide application causes runoff from agricultural areas into coastal estuaries. The levels of TAA were found to be elevated at low herbicide exposures. TAA and DAA may serve as valuable biomarkers of herbicide exposure in estuarine plants because the significant loss of TAA and rise in DAA were seen at increasing herbicide exposure concentrations (Bizily et al, 1999). These wetland plants may be able to survive some herbicide components of agricultural runoff in regions where they were utilized for phytoremediation due to the stimulation of ascorbic acid synthesis at low herbicide concentrations. In terms of economics, herbicides are crucial to American agriculture. The once labor-intensive production of corn (*Zea mays*), cotton (*Gossypium hirsutum*), soybeans (*Glycine max*), and other crops have become more productive because of chemical weed control. In order to provide crop protection throughout the entire growing season, agricultural herbicides are designed to be persistent. Herbicide use has improved agricultural productivity, yet there is a risk to the environment as a result. Herbicides may harm human health and natural ecosystems and disturb physiological processes in non-target plants. Some regions are frequently wooded zones with varied widths between agricultural land and surface waters in agricultural watersheds. They have high primary productivity, a seasonal flooding regime, and dense vegetation with species that can withstand floods. Their soils are often hydric, low pH, abundant in organic matter, with a thin oxidized surface layer that is covered by deeper reduced layers. The ability of riparian wetlands to retain and aid in the breakdown of resistant chemicals is an advantage that is frequently overlooked. The chemical conditions brought on by recurrent flooding and plant responses to these environments form the basis for this trait. A build-up of organic debris and partially oxidized zones around plant roots is common. The end result is a buffer zone that can delay herbicide migration and create conditions for resistant biological and chemical breakdown. Sometimes riparian wetlands are used for logging or field development. High-yield croplands can form from the conversion of wetlands to agriculture for a number of seasons, but this is followed by a slow drop in inherent productivity as the soils return to a nutrient-poor state. These areas have additional benefits when left in their natural state, such as protecting natural streams from the runoff of agricultural pollutants, runoff of silt, providing hydrology flood control, and maintaining wildlife habitat.

The percentage of water that exits the application site with surface flow is related to the protective effect of riparian wetlands. Herbicides that are not water-soluble leave behind sediment linked with erosion at the application site. Either the silt will flow into the drainage basin or it will settle away from the source of the flow of water. Riparian wetlands have hydrological advantages. By reducing the speed of stormwater before it reaches the stream, a riparian wetland along the edge can serve as a sediment filter. Topography or other natural obstacles, like vegetation, which can impede water flow rates, can exhibit this impact. When sediment containing herbicide is dumped in a wetland, the herbicide may escape using the same methods as it might in an agricultural field. However, the habitat in a marsh is significantly dissimilar from one in a field. The health of aquatic ecosystems and people are both at risk when pesticides are found in surface water. Pesticide contamination of surface waterways exposes unintended microorganisms, plants, and animals to chemicals that could be harmful to certain organisms or biotic ecosystems. Insecticides are particularly effective against insects and other arthropods, whereas herbicides may prevent the growth of aquatic plants. Long-term consumption of low levels of pesticides through drinking water is the main cause for concern when it comes to human exposure to pesticide-contaminated waters. Filtration, chlorination, softening, and recarbonation are common methods for treating water, but they don't do much to lower the level of pesticides in drinking water.

The concentration of pesticides in the drinking water will be similar to the quantities found in the surface water or groundwater source in places where water treatment facilities lack modern treatment procedures. The utilization of artificial wetlands and ponds with macrophyte culture for treating wastewater is currently being explored. The effectiveness and affordability of aquatic plant-based water treatment systems have been demonstrated in the improvement of wastewater effluent quality. Xenobiotic chemicals from polluted waterways and sediments have been removed and biotransformed by aquatic plants and the microbiota that they are linked with.

The biodegradation of surfactants is accelerated by the microbiota of *Typha latifolia L.* and *L. minor*. Pentachlorophenol was removed and degraded by curly leaf pondweed (*Potamageton crispus L.*), common duckweed (*L. minor*), and their epiphytic microbes. Additionally, Lemna and *Spirodela* spp. have been shown to accumulate metals like aluminum, cadmium, copper, lead, and mercury from aqueous solutions. Through plant absorption and microbial rhizosphere breakdown, vegetation can help soil ecosystems remove contaminants and organic chemicals generated by humans.

The area of soil that is affected by roots is known as the rhizosphere. Exudates and mucilage from plant roots are rich in energy and support vast and varied populations of microorganisms. The rhizosphere microbial variety and biomass make it superior for the breakdown of organic contaminants. Industrial chemicals including trichloroethylene, polycyclic aromatic hydrocarbons, and petroleum degrade more quickly in rhizosphere soil than in root-free soil. Plants may absorb pollutants as part of their transpiration process in addition to an accelerated breakdown in the rhizosphere. By boosting breakdown and uptake into tissues and limiting migration to surface waters and groundwater aquifers, vegetation may be crucial in cleaning up contaminated environments and preventing additional contamination. At petrochemical manufacturing locations, clay soils that have been contaminated with PAH are frequently found. Engineered remedial procedures that rely on excavation followed by thermal or chemical treatment frequently come at a high expense and pose a greater danger to public health since they release toxins into the atmosphere.

5.11 PAHS

PAHs can be metabolized by pure enzymes from cultures of bacteria, fungi, and algae. The availability of nutrients and electron acceptors limits the amount of microbial activity in the natural environment. The desorption kinetics, the population of PAH-degrading bacteria, and the competitive reaction for electron acceptor usage further limit the rate of PAH breakdown (AL Sbani et al., 2021). By integrating oxygen atoms into the ring structure and producing dihydrodiols via oxygenases, microorganisms normally break down PAHs aerobically. With the help of aromatic ring cleavage and subsequent oxidation, the derivative is further mineralized. Under oxic conditions as opposed to anoxic ones, PAH biodegradation rates are much higher. Microorganisms can break down low-molecular-weight PAHs containing two or three condensed benzene rings in an aerobic and denitrifying environment.

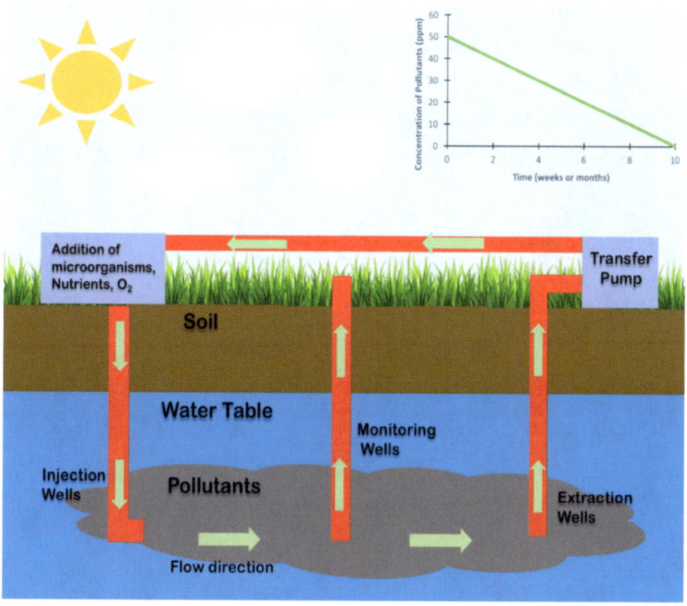

Figure 5.6. In situ bioremediation.

Source: *https://upload.wikimedia.org/wikipedia/commons/4/49/In_Situ_Bioremediation.png*

However, minimal biodegradation takes place in methanogenic or sulfate-reducing conditions. High-molecular-weight PAHs containing four or more condensed benzene rings are stated to be theoretically nondegradable at low redox potentials. Heterotrophic bacteria catabolize organic compounds; nevertheless, many of them are unable to decompose PAHs. The capacity for PAH breakdown by microorganisms is not indicated by the total heterotrophic population. In microbial populations, prolonged exposure to chemical toxins can lead to adaptations that increase toxicity resistance or improve the ability to use toxins as substrates for metabolism or co-metabolism. Chronic PAH exposure has been linked to enhanced PAH breakdown and microbial adaptations, although the mechanisms underlying these adaptations are yet unknown.

Clay soil affected by PAHs presents a problem for *in situ* bioremediation (Figure 5.6). The amount of PAHs that microorganisms can access may be constrained by the high adsorption capacity of clay soils. Microbial activities may be hampered by the poor flux of nutrients and electron acceptors in soil with low permeability. A designed process might speed up biodegradation, but it would be technically and financially impractical to distribute electron acceptors, substrates, nutrients, and enzymes to various microsites. However, it is known that plant roots carry out similar tasks. Microbial communities linked to plant roots in the rhizosphere constitute a diverse and dynamic milieu that has significant potential for detoxifying organic pollutants.

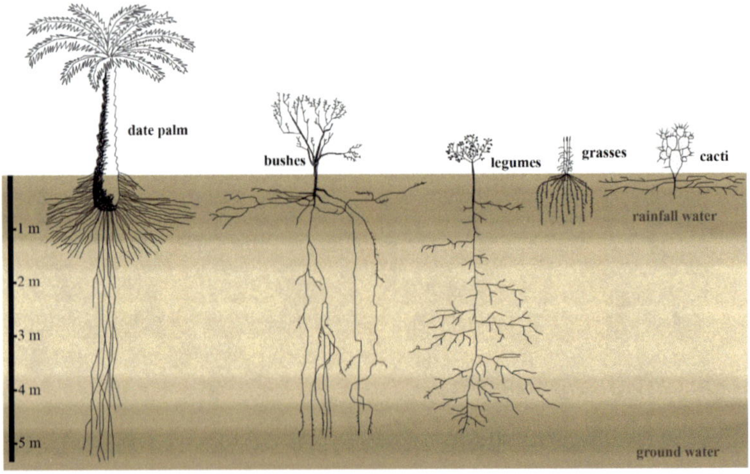

Figure 5.7. Deep fibrous root systems.

Source: *Kirschner GK, Xiao TT, Blilou I. Rooting in the Desert: A Developmental Overview on Desert Plants. Genes. 2021; 12(5):709. https://doi.org/10.3390/genes12050709*

Deep fibrous root systems (Figure 5.7) may enhance soil aeration by eliminating water through transpiration and changing the structure of clay soil through agglomeration. Organic matter in the soil is increased when plants discharge photosynthesized material into it through exudation and shedding of dead root cells. A large microbial community is supported by root exudates like carbohydrates, amino acids, etc. in the rhizosphere, which may promote the mineralization, polymerization, and/or breakdown of organic toxins. The group of fungi and bacteria found in the rhizosphere may be capable of a wide range of metabolic processes. The metabolism of high molecular weight PAHs, which bacteria and fungi cannot utilize as their only carbon source, may be supported by some root exudates. Furthermore, the higher soil organic content in rhizosphere soil may change the sorption, bioavailability, and leaching of toxicants.

5.12 CHEMICAL PROCESSES IN MOLECULAR BREAKDOWN

Different chemical processes are used by plants to break down complex aromatic compounds into simpler derivatives. There have been reports of aromatic ring-cleavage reactions in plant tissues that completely catabolize the aromatic nuclei into carbon dioxide. In plant tissues, the five-ring PAH benzopyrene can be converted to oxygenated derivatives. Despite the fact that some of these derivatives are more poisonous than the parent chemicals, they seem to be polymerized into the insoluble plant lignin portion, which may be a crucial detoxifying process. Benzopyrene was absorbed into organic acids, such as amino acids, in plant seedlings (Rentz, Alvarez, and Schnoor, 2005). With a variety of plants, complete benzopyrene degradation to carbon dioxide was also noticed. A healthy grass cover preserves soil and significantly lowers the danger of exposure to toxins in the soil. In general, soil-associated PAHs do not present a risk unless soil particles are directly absorbed, inhaled, or come into contact with skin. PAHs are poorly mobile because they are hydrophobic. With water flux, high-molecular-weight PAHs are essentially stationary. Grass reduces the risk to human health and the environment as soon as a dense grass ground cover develops, even if PAH breakdown, mineralization, and polymerization in soil may take some time. Grasses lessen the possibility of pollutant entry into groundwater and surface discharge. Due to animals grazing grasses, there may be a risk from plant bioconcentration of carcinogenic PAHs through food chains. Many plant roots and shoots have the capacity to bioconcentrate toxins found in the soil, including pesticides and petroleum-based substances. Due to PAHs' hydrophobic nature, they are primarily collected in the roots and do not spread to plant shoots. Low redox potential, desorption kinetics, a lack of microbial community, and a lack of nutrients may all be factors limiting PAH breakdown in clay soil. The biodegradation of PAHs may be aided by a root zone system. Grasses significantly lower the chance of coming into contact with contaminated soils and may eventually turn PAHs into harmless compounds. Although PAHs are not translocated to plant shoots, they may be bioconcentrated to plant roots.

CHAPTER 6

Metal Contaminated Soils

CONTENTS

6.1 INTRODUCTION

A number of environmental regulations were passed in the United States as a response to the recognition of widespread environmental contamination, including soil and groundwater. Most states have their own laws, rules, and regulations regulating the remediation of soils when those soils directly or indirectly endanger human health and, in many cases, the environment. These laws are based on federal statutes. Systematic soil sampling, site investigations, reporting, data compilation, and registries are often required to determine the number of suspected or confirmed contaminated soil sites in a specific area. There are several countries where only minimal data is accessible. Estimates are subject to variation, and they may not be accurate or reflect levels of soil contamination that actually exist, even in highly industrialized nations. It is difficult to determine the precise global scope of soil pollution.

Figure 6.1. Soil contamination: contaminated soil in the Linares mining district: (**a**) general image of the area and (**b**) detailed image of the soil.

Source: *Terrones-Saeta JM, Suárez-Macías J, Bernardo-Sánchez A, Álvarez de Prado L, Menéndez Fernández M, Corpas-Iglesias FA. Treatment of Soil Contaminated by Mining Activities to Prevent Contamination by Encapsulation in Ceramic Construction Materials. Materials. 2021; 14(22):6740. https://doi.org/10.3390/ma14226740*

Over the past two decades, both the number of governmental organizations mandating and supervising soil remediation and the number of consulting experts carrying out soil remediation activities have increased. Many of these experts may be skilled at remediation, but they might not completely comprehend why they remediate. These comprise both conventional remediation technologies and non-conventional techniques (Mahar et al., 2016).

6.2 SOIL CONTAMINATION

Soil contamination (Figure 6.1) happens when a solid or liquid substance interacts with soils and becomes chemically or physically bonded to soil particles or lodged in the spaces between them, posing a real risk to human health or the environment. For instance, adding water or fertilizers to the soil at the appropriate rates of nutrients would not produce a contaminated mixture. Contamination could happen if the fertilizer was administered excessively. Chemicals can injure people, plants, and animals, leak toxins into groundwater at unacceptable levels, and severely degrade soil resources when they are present in significant quantities, which is when they become soil contaminants. When soil is found to be excessively contaminated, it can be cleaned up by soil excavation, treatment, and/or disposal, in situ soil treatment, soil containment to stop soil contaminant movement or limiting exposure to the contaminated soil. Due to the increased interest in reusing brownfields, existing contaminated soil sites are being closely examined. The United Kingdom declared in 1998 that 60 percent of all new residences and commercial buildings would be constructed on Brownfield sites (Batty and Anslow, 2008).

The level of scrutiny used to address these issues will change depending on the political and regulatory context of a particular place. Risk assessment for human health is a method, and it is difficult, time-consuming, and sometimes expensive. In accordance with the USEPA's standard risk assessment paradigm, risk assessors determine the potential risks or hazards posed by these chemicals to human or ecological receptors after evaluating the hazards of the chemicals in contaminated soil, determining the environmental fate and transport of these chemicals. Soil clean-up or remediation may be required if certain soil concentrations, either separately or collectively, are over permissible limits. Comparing soil contamination levels to regional soil background levels is an important technique. Soil clean-up or remediation may be required when soil contamination levels are higher. To establish cleaning levels and how much residual chemical pollution will remain on a site once remediation activities are finished, government risk managers employ the risk management process.

Risk assessments do result in numerical risk levels or risk-based clean-up concentrations, but they do not take into account a large range of additional criteria that must be considered when making a final risk management decision. Political considerations, social considerations, and economic considerations are significant factors in the decision-making process. This is necessary in order to make an informed remediation decision. In order to make an open, consistent, and repeatable decision that will stand up to peer and legal review, the ideal risk manager's decision-making process integrates site-specific information, general information about legislation, laws, regulations, policies, and rules, regulatory guidelines, scientific data, engineering feasibility, and other information. Numerous anthropogenic activities may cause soil pollution. Soil contamination may result from ordinary industrial operations, air pollution, agricultural and forestry practices, and other factors. The Scottish Environmental Protection Agency lists a number of potential negative effects from these activities, including altered rates of nutrient cycling, altered rates of organic matter decomposition, reduced microbial biodiversity and biomass, loss of organic matter and subsequent harm to the physical structure of the soil, mobilization of potentially toxic elements, loss of ecosystem functions in the soil, and detrimental effects on the water and air (Pulford and Watson, 2003). Some of the laws demanded that contaminated soils be looked into and remedied to a point where they wouldn't be a risk to human or environmental health. The form and degree of soil pollution are poorly understood due to a lack of systematic soil sampling, site inspections, reporting, and data compilation. It is obvious that major soil contamination is a global issue. Understanding the exact extent of soil pollution in a given location, region, country, or continent is challenging due to the inconsistent reporting of contaminated soils. Although the degree of soil contamination globally is unknown, this is a problem that affects all continents and is likely to get worse as more data is gathered. Resource-intensive activities like remediation are common. As a result, while choosing which remediation option to use for a particular site, cost and remediation time become crucial considerations. It is expected that some more regions will be remedied using bioremediation and phytoremediation, which are important and affordable alternatives. The plant species of both naturally occurring and human-influenced metalliferous soils are being explored more and more, both as independent ecosystems and in search of accumulator plants that may be useful for various types of phytoextraction. More than 440 hyperaccumulator species have been identified so far, with three-quarters of these being nickel accumulators from the widespread occurrences of ultramafic soils rich in nickel that are found all over the world (Reeves, 2006).

6.3 RESEARCH

For a long time, the majority of research was focused on crops that were significant to both humans and animals, and it was constrained by the analytical methods that were available for the study of major elements like calcium, potassium, magnesium, nitrogen, and phosphorus. More information on minor and trace elements has become available during the 20th century as a result of the development of atomic spectroscopic methods like flame emission, atomic absorption, and plasma emission, as well as the improvement of the specificity and sensitivity of analytical techniques. In addition to the aforementioned elements, other elements that have been demonstrated to be crucial for plant growth include B, Cl, Cu, Fe, Mn, Mo, and Zn may now be monitored. Although they may not be necessary, some other elements, including Ni, Se, Co, and V, have either been demonstrated to be required for specific species of organisms or at the very least to have a favorable impact on some plants. The effects on plants of metalliferous soils, which typically contain these metals as minor elements (Mn) or trace elements (Zn, Cu, Ni, Cr, Pb, As, Co, Se, Cd), can vary greatly. There are extreme cases when soils are toxic to nearly all higher plant species, but under other conditions, a distinctive flora of metal-tolerant species may arise. It is usual to see a little higher than normal concentration of the elements that the soil is enriched with, suggesting that the metal-rich environment has only a minor impact on the elemental composition of the majority of metal-tolerant organisms.

6.4 PLANT SPECIES

A few plant species have adapted to only occupy the ecological niche that the metalliferous environment offers. These species have been employed as geobotanical indicators in mineral exploration where it has been demonstrated that they are indigenous to regions with clearly defined soil composition. The ability to accumulate very high concentrations of specific metals, much above normal physiological requirements, and levels far exceeding those seen in the majority of other species tolerant of metalliferous soils, is another characteristic of some metal-tolerant plant species (Dunn et al, 1996). A plant may also take up high levels of metal under artificial conditions, such as through significant metal-salt supplements to an experimental soil or a nutrient solution. The forced uptake of extra metals from fertilizer solutions frequently results in plant death or other delays in the full development of the plant. Even though they might be useful for some phytoremediation procedures, such trials frequently have little bearing on the behavior of naturally existing metallophyte populations. However, there is one more situation that needs to be taken into account: if given the chance on a native metalliferous soil, a plant species that is never discovered to be a hyperaccumulator may do so. A species like this might be referred to as a facultative hyperaccumulator, especially if it can be demonstrated that it successfully completes its life cycle on such a soil sample. It is frequently the case while looking for hyperaccumulator plants that the proper soil samples have not been obtained with the plant material, making it impossible to apply a specific plant/soil concentration ratio.

6.5 METALLIFEROUS SOILS

Metalliferous soils (Figure 6.2) come in a wide variety of both naturally occurring and artificially produced varieties. The most significant types of naturally metalliferous soil, which may support some hyperaccumulator plant species, are serpentine soils, which are derived from Fe and Mg-rich ultramafic rocks and are found throughout the world, soils enriched in Pb and Zn, which typically contain elevated Cd as well and occasionally carry high concentrations of As and/or Cu and Co containing soils. Serpentine soils are widespread in many nations, especially those that have experienced intense tectonic activity. They are a significant component of the geology of places like Japan, the Philippines, parts of Indonesia, California, Turkey, Greece, the Balkan lands, parts of the European Alps, central Brazil, Cuba, and Puerto Rico. Particularly from the Cretaceous shales of the midwestern United States, notable Se-rich soils have developed. Both of these soil types frequently host distinctive metal-tolerant plants, some of which may be hyperaccumulators of Ni and Se, respectively. Many regions around the world have restricted access to metalliferous soils produced by human activities, often with distinctive floras. These include soils generated on the remains of the mining and smelting of ore containing the metals Ni, Co, Cr, Cu, Pb, Zn, Cd, and As. Other human uses of metals or metal compounds have also resulted in local soil contamination, which now requires some sort of clean-up (Ahemad, 2015).

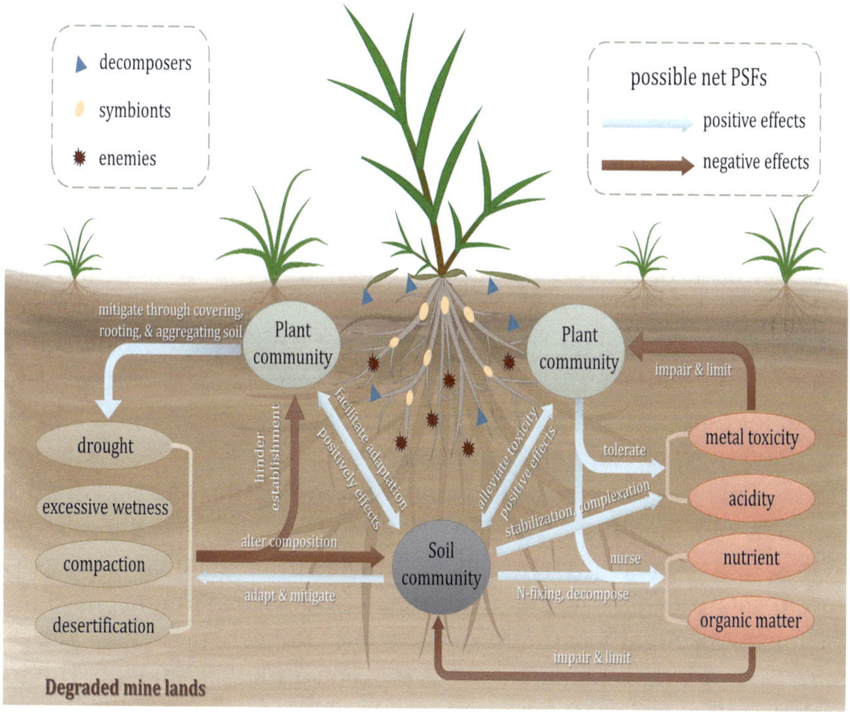

Figure 6.2 Metalliferous soils

Source: *Zhu S-C, Zheng H-X, Liu W-S, Liu C, Guo M-N, Huot H, Morel JL, Qiu R-L, Chao Y and Tang Y-T (2022) Plant-Soil Feedbacks for the Restoration of Degraded Mine Lands: A Review. Front. Microbiol. 12:751794. doi: 10.3389/fmicb.2021.751794*

The ppb (ng/mL) range of sensitivity for the majority of the elements is more than sufficient for analyses of the majority of the elements in contaminated soils and accumulator plants. Metal vapors and very fine dust may deposit on plant material during the sampling of vegetation close to smelters and cause contamination that is much more concentrated than that produced by soil dust. Special care must be taken during the washing procedures since any contamination of plant material from such sources is likely to flow readily into the solution under the conditions employed for plant analysis. It is possible to eliminate particle deposits by briefly agitating the plant material for a few minutes in water with detergent, but this cannot always be guaranteed. Where this seems to be an issue, leaf material can be suspended in an ultrasonic bath with a non-aqueous inert solvent to help remove fine particles off the leaf surfaces. The attention paid to analytical studies on ultramafic floras may have contributed to the high number of Ni hyperaccumulators in comparison to those of other elements, but the area's size and the sheer quantity of tracts of nickeliferous soils are almost definitely also major factors. In nations like New Caledonia, Cuba, and Turkey, continuous ultramafic expanses of tens or even hundreds of square kilometers can be found. This has provided an opportunity for the evolution of a distinctive flora with many endemic species, including some that have developed Ni accumulation, where such areas have been continuously available for plant colonization for many millions of years. In the case of the Cuban serpentines, the influence of soil exposure age appears to be particularly important. Other species, such as the western and central European *Arabidopsis halleri* have been shown to contain more than 10,000 mg/kg Zn. Lead is present in the majority of soils and rocks at levels below 50 mg/kg, and it normally moves through soils and into vegetation with mobility of only around 10 mg/kg Pb on average soils. Even in above-ground plant sections where amounts of 1–10 mg/kg are reported, some of this probably results from other types of environmental and lab contamination. Significant translocation to the upper sections of the plant is rare when Pb does penetrate the plant roots in larger amounts from Pb-enriched soils (McGrath, Zhao, and Lombi, 2001).

Copper and cobalt levels in plants typically vary from 5 to 25 mg/kg and 0.03 to 2.32 mg/kg, respectively. *Nyssa sylvatica*, also known as tupelo or black gum, is a remarkable plant that can accumulate up to 845 mg/kg of cobalt from common soils. But even on soils rich in cobalt, like those made from ultramafic rocks, plant Co hardly ever rises above 20 mg/kg. Even in the presence of high soil Cu, plant Cu concentrations are kept within a small range, and Cu concentrations above 100 mg/kg are infrequently observed in thoroughly washed plant leaves. Manganese levels in plant dry matter typically vary from 20 to 500 mg/kg, occasionally exceeding 1000 mg/kg in plants growing on typical soils. When the leaf concentration is below 10 mg/kg, deficiency symptoms can frequently be identified (Zitka et al., 2013). In contrast, toxic levels, which vary depending on the species, are between 1000 and 12,000 mg/kg. It has been shown that manganese complexes play a crucial function in the oxidation of water to oxygen as well as being an essential ingredient that activates some of the tricarboxylic acid cycle enzymes. The manganese concentrations of plants on soils developed over Mn deposits, particularly the oxide minerals pyrolusite and psilomelane, have received relatively little attention. On soils with manganese mineralization and soils with lower Mn concentrations, some species have been discovered with 1000–5000 mg/kg Mn. Selenium is a crucial element for both human and animal health, with an astonishingly small window between the amounts needed to prevent deficiency disorders and those that cause toxicity symptoms. The Se content of soils typically ranges from 0.01-2 mg/kg, but it can go much higher than this, with concentrations of several hundred mg/kg in soils derived from specific cretaceous shales or where the soil has become Se-enriched through the disposal of materials like fly-ash or the evaporation of irrigation water. Se-rich soils naturally occur in parts of the western United States, as well as in Queensland (Australia), Ireland, Colombia, and Venezuela. Zn, Ni, Co, Mn, and Cd can be accumulated to high concentrations in the leaves when these elements were introduced as low-concentration supplements nutrient solution. Other substances like Pb, Cu, Fe, Al, and Cr were heavily concentrated in the roots but were not easily transferred to the shoots. The behavior of some populations of *T. caerulescens* in the wild, particularly those on Zn mine and smelter sites in south-central regions of France, has also demonstrated that this species is capable of accumulating Cd. These populations are currently gaining attention due to their potential for Cd phytoremediation.

6.6 CHARACTERISTICS OF *T. CAERULESCENS* AND OTHER SPECIES

Although the characteristics of *T. caerulescens* have received a lot of attention recently, more in-depth research on the closely related *T. brachypetalum* and other species like *T. goesingense, T. cepaeifolium, T. alpinum,* and *T. ochroleucum* is justified. DNA study provides significant evidence that this species is distinct from the European species of the same name, which is not a metallophyte. More research is essential to better understand the intra-species connections and assess the phytoremediation capacity of other populations of the North American species. Whether they are hyperaccumulators or not, many plant species found on serpentine soils are rare and endemic species. Alyssum, Bornmuellera, Thlaspi, and Centaurea are a few Ni-accumulating species that are only known from two or three sites or collections in Turkey. Even some of the more common metal hyperaccumulators might be at risk on a local level. For example, populations of the Zn-accumulating *Thlaspi caerulescens* are abundant in some places but in danger of extinction in others. *Arabidopsis halleri* populations can vary greatly in size and density. It is crucial to protect all populations of these intermittently distributed species and to ensure that they are kept in good condition because there can be significant variation in the metal-accumulating characteristics within and across isolated populations of such species (Roger, 2001).

An effort is now required to enhance our field knowledge of metal-tolerant species, and hyperaccumulators in particular, as they are a valuable and potentially useful biological resource. These plants are found throughout the world, and it is important to update this knowledge. This comprises areas of waste such as from the mining of Cu, Zn, and Pb in many different countries as well as expanses of serpentine soil, particularly in the tropics. The development of strategies for protecting endangered populations and rare species will benefit from this information. The metal-hyperaccumulating plants are a unique and significant biological resource with a lot of promise for utilization in various soil remediation methods. In order to identify new accumulator species and better understand their distribution, there is an urgent need for more investigation of the metalliferous soils. Some issues need to be resolved regarding how scientific and commercially oriented work can proceed while promising benefits to the people whose territory the resource is found in nature, as well as conservation measures that need to be put in place to ensure that rare hyperaccumulator species are not lost due to any of the threats. Lead and arsenic have never demonstrated much mobility. Out of all the elements studied, arsenic had the lowest amount in plant tissues. Zinc and cadmium were easier for plants to absorb and more mobile. Although their transfer to aboveground biomass was quick and bulk roots occasionally had lower element concentrations than leaves and stems, secondary roots often contained their maximum levels. The high concentration of zinc in seeds, along with the fact that seeds and other edible vegetable portions were well protected from Cd buildup, demonstrated the importance of zinc for plants.

Plants, which cover the nearly solid surface of the planet, have a variety of uses, including providing food for people and other animals and serving as construction materials, generating energy, primarily heat, and serving a number of industrial functions. Plant species and genotypes have a significant impact on the natural sensitivity or tolerance of plants to accumulate metals. The majority of agricultural plants that respond more or less linearly to an increase in the availability of trace elements in the soil are considered indicators. They are wheat, oats, and maize crops. Plants that accumulate more elements in their tissues as they grow in the soil are known as accumulators. Numerous plants from the Brassicaceae (mustard) and Compositae (lettuce, spinach) families are accumulators. Extreme accumulators, or hyperaccumulators, can even live on contaminated soils and collect high levels of trace elements. These groups mostly display the uses of plants, implying the presence of numerous plant species in each. When compared to *Thlaspi caerulescens*, a zinc hyperaccumulator that produces ten times less shoot biomass, *Brassica* spp. appears to be more effective at removing zinc from the contaminated soil. Not only do variations in the way that elements accumulate in plants and their yield of biomass affect the overall amount of biomass but also the unique characteristics and mobility of each element can have a big impact on how it is absorbed and distributed in plant tissues. With a rise in the concentration of Cd in the soil solution, Cd content in plant tissues increased.

6.7 TRACE ELEMENTS

Cd, Mn, Mo, and Zn were easily transferred to the plant tops whereas Pb, Hg, and Ni were the least frequently translocated. The mobility of trace elements in plant tissues and their overall concentration in plants did not match the element content and variations in the soil solution. Within a single plant species, the composition of individual elements can vary by two or three orders. Researchers categorized the elemental content into three categories: high, common, and deficient in the case of micronutrients. The accumulation of elements in plants was not solely dependent on the type of plant and how the components behaved. There are large variations in the total amount of soil elements, and they can be drastically altered by contaminated soil. Individual element release is influenced by soil characteristics, including pH, CEC, clay, oxide, and organic matter levels. Transfer coefficient (TC), was released. Total element concentration in plant tissues was divided by total soil content to obtain TC. The highest and lowest transfer coefficients for every element varied. Distinct mobility of trace elements was confirmed by TC. Except for maize and peas, plants typically introduced greater TC for Cd and low TC for Cr and primarily Pb. The remediation factor (RC), which compares the quantity of an element removed by above-ground biomass to the total amount of an element contained in the soil's top layer is important.

Due to their geogenic and anthropogenic origins, agricultural soils can have significantly different soil elements. For agricultural and horticultural production, several soils with high metal levels are used, which affects the accumulation of elements in plant tissues. Growing Cd concentrations at Luvisols caused increased Cd to accumulate in the green biomass of the plants like oat and maize. A two-fold increase in soil Cd content did not result in any appreciable changes at any of the plant parts, confirming the importance of soil characteristics for element accumulation. The largest rise in Cd content was found in maize biomass, which increased practically linearly in response to an increase in Cd soil concentration. The lowest reaction was found in grains at both Luvisols and contaminated Chernozem. Barley grain demonstrated a strong defence mechanism, and its Cd content was only seven times greater than that of soils, which varied by a factor of twenty. In comparison to white sweet clover, mallow and safflower often introduced higher accumulation potential. Red clover consistently has a few soil pollutants. In many affluent nations throughout the world, the replacement of agricultural crops with energy crops that produce biomass as a sustainable energy source is becoming a reality. When an appropriate accumulation of components and biomass production can occur, non-food or food biomass production can also be useful for the phytoextraction of pollutants from the soil. Willows are also suitable for environments with high metal concentrations and tolerance (Tlustoš et al, 2006).

6.8 WEED PLANTS

Weed plants have a very excellent ability to survive. Without a direct relationship to the temperature and soil conditions, weed species can potentially produce significant amounts of biomass. *Artemisia vulgaris* and *Lathyrus pratensis* are good bioindicators of element contamination and accumulation among the variety of weeds evaluated. Plants directly reacted to mild increases in the soil's element content by accumulating elements. Under conditions of high soil element content, plants lost their ability to accumulate Cd and Zn, and the increase in soil element content outpaced the growth of plant content. Extremely high Zn soil contents slowed down plant growth, while excessive levels of other elements had little effect on phytotoxicity. Consumable portions of vegetables and cereal seeds were both well-protected and did not accumulate any harmful substances. Due to root degradation and element release, different root development and accumulation capacities for heavy metals had an impact on the planting of the subsequent crops even though they did not directly harm the first crop. *Arabidopsis halleri* cultivated in a contaminated environment was confirmed to have the maximum Cd and Zn accumulation capacity. Some species, like *Trifolium pratense* (Figure 6.3) or *Rumex acetosa*, showed a limited propensity to take up elements, whereas other species did not accumulate enough elements in their green tissues (Schwitzguébel et al, 2011).

Figure 6.3. Trifolium pretense.

Source: *https://upload.wikimedia.org/wikipedia/commons/thumb/a/ae/Trifolium_pratense_-_Keila2.jpg/800px-Trifolium_pratense_-_Keila2.jpg*

6.9 GM PLANTS

GM plants produce pharmaceutically significant compounds like antibodies or enzymes The use of plants as heterologous expression systems, such as those for the synthesis of antigens of mammalian diseases, steadily became a new area of GM plant technology in the 1990s. Different therapeutically significant antigens, such as the surface antigen of hepatitis B or rabbit viral glycoprotein, were already expressed in transgenic plants. Genetic engineering is increasingly being used in plant biotechnology. Genes control ripening, fat composition, and the quality of seed proteins, whereas other transgenes affect crop output and quality. Research on tissue-specific promoters will enable genetic engineering to develop new types of GM plants that can produce the necessary products. The growing use of transgenic plants generated significant debate over the past twenty years. This is especially relevant to industrially produced crops used as sources of nutrition for people. Unfortunately, the majority of objections to the use of GMOs lack any scientific foundation, notably those made by activists in various green movements. That explains why even good arguments are frequently rejected. There is no contradicting the need for caution when handling, cultivating, and particularly using genetically modified organisms. It is also essential to assess the hazards associated with their discharge into the environment or onto the market. Many nations have enough laws governing food manufacturing, including the requirement to label any food item containing transgenic DNA beyond a certain threshold. The methods for detecting the presence of transgenes are frequently evolving and improving.

6.10 PLANT BIOTECHNOLOGY

The plant biotechnology revolution gave rise to two key discoveries. The first was the extraordinary capacity of any single plant cell to develop into a complete plant. Similar in importance was the description of the Ti plasmid of soil bacterium *Agrobacterium tumefaciens* and the Ri plasmid of *A. rhizogenes*, which are currently the most effective vectors for gene transfer into plants. One of the primary methods for introducing foreign genes into the plant genome is by the use of agrobacterium gene transfer techniques, however, even this widely used technology has certain inherent limitations. Dicotyledonous plants can be successfully transformed using agrobacterium plasmids as vectors, but effective uses on monocotyledonous plants have also been observed. Viral vectors are a different method of natural gene transfer, although they only produce temporary changes. This approach of transferring genetic material into plant cells is based on the inherent capacity of soil bacteria *Agrobacterium tumefaciens* and *A. rhizogenes* to infect higher plants and introduce their genes into them. These genes can cause morphological changes in affected plants and are located in a section of a big plasmid known as Ti (tumor inducing) or Ri (rhizome inducing). T-DNA is the portion of Ti or Ri plasmids that is transmitted to plants. Plasmid Ti typically ranges in size between 150000 and 200000 base pairs or roughly 3% of the chromosome of *A. tumefaciens*. Two major components make up Plasmid Ti. The virulence region, contains the genes required for the functions permitting T-DNA transfer to plant cells and their integration into the plant genome. T-DNA, which enters the plant cell but lacks the genes for its own integration. Other sections of the genome, some of which are also localized in the bacterial chromosome, are also required for the operation of the plasmid in bacterial cells and the interaction between bacteria and plant cells. Genes for the manufacture of plant hormones, auxins and cytokinins, are two sets of genes that the T-DNA of wild-type Agrobacterium inserts into the plant genome. They alter the shape of the tissue and prompt plant cells to differentiate, leading to the production of undifferentiated tumors or so-called hairy roots (Figure 6.4).

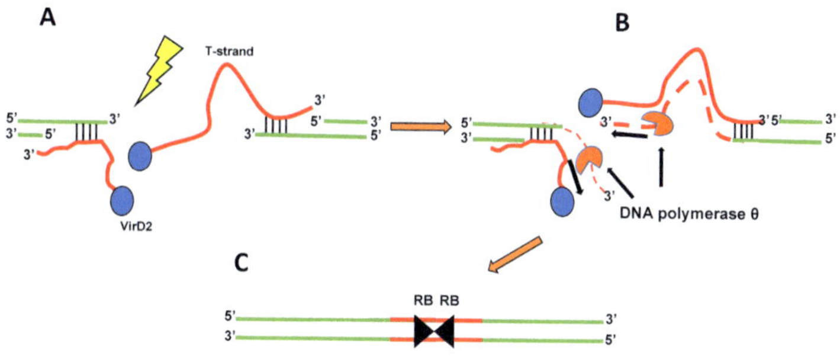

Figure 6.4 Agrobacterium T-DNA integration.

Source: *Gelvin SB. Plant DNA Repair and Agrobacterium T−DNA Integration. International Journal of Molecular Sciences. 2021; 22(16):8458. https://doi.org/10.3390/ijms22168458*

6.11 OPIATES

Opiates serve as a source of carbon, nitrogen, and energy. Currently, strains containing modified Ti plasmids are employed for genetic manipulations instead of the native wild-type Agrobacterium strains. The T-DNA with border sequences was inserted into smaller plasmids derived from pBR 322, which are now commercially accessible, as the large size of the native Ti-plasmid causes handling issues. Unique restriction sites are present in the T-DNA region of these newly created plasmids, enabling the insertion of constructs ready for transfer into the plant genome. A genetic construct that is meant to be introduced into plants must have the proper promoter region, the sequence coding the required gene, and the termination signal. These promoter regions are typically constitutive promoters from the cauliflower mosaic virus CaMV 35S that control ubiquinone expression in corn, or other organ or tissue-specific promoters active in plants. Currently, a lot of work is put towards isolating new promoters that can be induced, for example, by an environmental influence like the presence of heavy metals. The ability to establish cell fusion between cells of closely related species as well as very distinct plant species under specific conditions makes protoplasts an effective instrument for gene manipulation. This enables the production of somatic hybrids. You can obtain protoplasts from plant tissues, calluses, or suspension-cultured cells. Callus tissue can form following protoplast fusion and transfer to a nutritional medium containing cytokinins and auxins in the correct ratio. This can be accomplished approximately 5 to 10 days after the initiation of cell division and the construction of a new cell wall, following which cell differentiation takes place and an embryo is generated. Such an embryo can grow into a fully developed plant with modified genetic machinery. Ballistic techniques, which enable the transfer of genes into cells and entire intact plants, are further helpful tools. For species for which the use of Agrobacterium failed, relatively innovative strategies are applied. The method can be successfully applied to a variety of additional tasks, such as direct genome transformation of organelles or rapid delivery of genetic material into cells without causing tissue damage. Microinjection, which transfers DNA into the plant cell while being controlled at a microscopic level, is another recognized form of gene transfer.

6.12 DNA INSERTION

DNA can be inserted into specific cells inside multicellular structures as well as into isolated protoplasts with defined subcellular compartments. This technique is the most effective, and it also works with plant tissues. The drawback is that just a small number of cells can be impacted by a single experiment. Due to its high efficacy, low copy number, intact transgenes, and ballistic procedures which enable transformation regardless of the target plant species, the majority of laboratories working in this field choose the agrobacterium transformation over the other ways. The main issue that could be solved by using transgenic plants more widely is the continually rising food consumption. The global population has grown by 90% over the past 40 years, yet food production has only expanded by 25% (Licht and Isebrands, 2005). Another issue is the annual crop losses brought on by insect or other pest attacks, despite the fact that the amount of agrochemicals like pesticides, insecticides, or herbicides used cannot be increased for ecological reasons. The primary goal of genetic engineering is the production of organisms that are resistant to pesticides, bacterial, fungal, or viral diseases, herbicides, and pests, as well as organisms that boost the nutritional value of the food that is produced. The introduction of herbicide and insect pest resistance can be regarded as the most significant. In the USA, glyphosate-resistant varieties of canola, cotton, and soy were brought to the market, and since 1996, their share of the overall yield has been steadily rising (Aslam and Gul, 2020).

The majority of all plant GMOs are insect-resistant species. These plants have genes for toxins made by the bacteria *Bacillus thuringiensis* in their DNA. The mining sector, industrial manufacturing, and various urban activities have all contributed significantly to environmental contamination over the previous two centuries. Without even mentioning the wars and other conflicts that are responsible for the development, usage, and storage of explosives and chemical warfare agents, agricultural practices, particularly those of the twentieth century, are to blame for the widespread pollution of soil, sediments, and water. Another significant issue is the air pollution caused by harmful volatile substances. The physico-chemical procedures are one option for removing pollutants from soil. Unfortunately, these methods are very expensive and frequently worsen environmental damage. Due to this, a significant amount of research in the last decade focused on the application of biological remediation techniques. One of them is phytoremediation, which is the use of plants to transmit, accumulate, and remove contaminants from the environment, or at the least to reduce their spread. It is a method that can be used to eliminate both inorganic and organic pollutants from soil, water, and the atmosphere.

The selection of plant species suitable for phytoremediation might be considered one of the primary goals of fundamental research in the field. Different plant species have different capacities for metabolizing or storing contaminants. A plant must be able to accumulate harmful metals, transport cations or oxyanions into the harvestable sections, or change them into less-toxic forms in order to remove heavy metals from the environment. It would be ideal if plants working with a microbial consortium could completely mineralize the molecule into non-toxic compounds during the clean-up of organic contaminants. The primary limiting element in the extensive use of plants is the lengthy period required for soil decontamination, particularly during the accumulation of heavy metals but also during the elimination of organic contaminants. Due to this, considerable effort is put into growing or genetically modifying plants to make them specifically suited for phytoremediation. More information regarding ways to boost the expression of genes or introduce bacterial or mammalian genes into plants in order to improve the natural capacity of plants to deal with xenobiotics has recently surfaced. The rates of metal absorption and storage are governed by the interplay between transport and chelating processes. The transport and storage forms of transition metals are still largely unknown, and many systems have not yet undergone analysis. Numerous attempts to produce transgenic plants with enhanced traits for phytoremediation have already been made.

6.13 TRANSGENIC PLANTS

Only a few transgenic plants have been evaluated on actual polluted soil. Mercury was once heavily used in the chemical, paper, and mining sectors. The wastewater from production plants was frequently discharged into adjacent water resources because there was little understanding of the toxicity of mercury and its potential impact on the environment. For instance, this led to two widespread intoxication incidents in Japan in the 1960s (Bhan and Sarkar, 2005). Clinical studies demonstrate that organic derivatives, such as the cation of methylmercury (CH_3Hg^+), are the species of mercury that accumulate in fish and fish eaters most frequently and are responsible for neurological symptoms. Methylmercury is a relatively uncommon waste; however, it can be produced when inorganic mercury is converted by microbes. The cation CH_3Hg^+ has higher biological availability than the salt of Hg(II) because it is less tightly bound in complexes with organic and inorganic ligands in sediments. It is also more soluble in lipids, which increases its toxicity. Only a small number of mercury-contaminated sites have been cleaned up due to the high costs associated with excavation and storage of hazardous materials. Research has recently focused on phytochelatins as one of the natural processes that enable plants to grow in environments that are contaminated with heavy metals. In the plant kingdom, phytochelatins are commonly used to store heavy metals. When plants are exposed to Cd or other heavy metals, phytochelatin synthase converts glutathione into these peptides. It has been reported that some plant species can accumulate significant levels of heavy metals. Hyperaccumulators have a much greater capacity to accumulate heavy metals than other species that are growing in the same environment at the same location. Unfortunately, hyperaccumulators are characterized by a relatively low biomass formation, which makes them unsuitable for phytoremediation. They are typically found in locations where metal ore with a high metal content is present. Nevertheless, efforts have been made to increase the accumulation of heavy metals in plants producing large amounts of biomass, but the mechanisms underlying hyperaccumulation are still poorly understood.

6.14 TRANSGENIC PLANTS FOR METAL UPTAKE

In order to boost the uptake of metals, transgenic plants were produced with higher levels of phytochelatin synthase or glutathione synthase, which both increased Cd accumulation. Transgenic plants with foreign genes for proteins transporting metal across membranes have also been produced.

Various metallothionein genes from mammals, yeast, insects, and humans were inserted into plants in some labs. This effort mainly increased the plant resistance to some heavy metals but did not yet result in an increase in accumulation. The performance of transgenic plants may be enhanced by improved bioavailability of metals brought on by changes in exudate, greater excretion of organic acids, or collaboration with rhizospheric bacteria. The development of phytosiderophores showed that their presence is ubiquitous in higher plants, it may be enhanced or alternatively produced by suitable bacteria. Cytochrome P_{450} is well known for its capacity to take part in the oxidation of a variety of substances, including several xenobiotics (Macek, Macková, and Káš, 2000). Cytochrome P_{450} is found in all living things, including microorganisms, plants, animals, and mammals. It participates in a wide range of chemical reactions, including aliphatic hydroxylation, epoxidation, dealkylation, dehalogenation, and other processes that result in inactivation and detoxification. A large number of these reactions are essential for effective bioremediation. Due to their abundance and powerful reduction and oxidation capacity, microbial P_{450} are particularly helpful for biodegradation processes. Recent research has demonstrated that cytochromes P_{450} can naturally break down substances including atrazine, thiocarbamate herbicides, and haloalkanes. The genes for oxidoreductase and cytochrome B5 were inserted to ensure the stability and activity of the enzyme.

CHAPTER 7

Perennial Grass and Pollution Management

CONTENTS

7.1 INTRODUCTION

According to several studies, certain plant species can absorb, metabolize and detoxify heavy metals or inorganic pollutants from contaminated soil and waste materials. This procedure is frequently incorporated into the holistic strategy used for ecological restoration projects on damaged lands. The plants that are best suited for this process have high growth potential, high pollutant accumulation potential in their roots and shoots, tolerance from the pollutants, and pollutant detoxification potential. The grass (Poaceae) family of plants has shown to have enormous potential in this area, as evidenced by the numerous. On the basis of their tolerance ability to metals or contaminants, these grasses are frequently categorized as accumulators and excluders. Accumulator plants have the capability to maintain metabolic stability in situations where metal concentrations in their photosynthetic sections are relatively greater than the maximum permitted limits. This is owing to the fact that these plants effectively combat the oxidative stress brought on by high metal buildup. While restricting the transport of pollutants up to the aerial portions, excluder grass plants either accumulate pollutants in their roots or transform them into complexes in the root zone. These roots are more tolerant of metals than their shoots, and following detoxification, the absorbed metals frequently collect inside the root cell vacuoles (Rylott and Bruce, 2009).

7.2 BIOAVAILABLE METALS

Around the soil particles, the bioavailable metals are divided between the solid and solution phases through chemical processes like adsorption, complexation, precipitation, and redox reactions. Through the root membranes, the roots absorb the metals in the solution phase. The steps of the metals' transport mechanism, which differ depending on the species, include metabolization, detoxification, and compartmentalization. Furthermore, grasses have complex root systems with numerous subterranean stems called rhizomes that link the soil and prevent erosion and contaminant leaching. The quick formation of a dense green cover over large areas of polluted land is also aided by these structures. Using grasses as the colonizers ultimately helps to create an attractive landscape. In particular, aromatic grasses have been demonstrated to be significantly more effective for phytoremediation of a variety of contaminants from wastelands, with benefits ranging from ecological to socio-economic importance. The genus Vetiveria sp. (Vetiver grass) and Cymbopogon sp. (Lemon grass) contain some commonly grown aromatic grasses. Essential oils are one of the products that aromatic grass produces, and they have the advantage of not being polluted even after the metal has been absorbed by the tissues (Banerjee et al, 2019). Numerous studies have also demonstrated that different abiotic stresses do not affect these plants' ability to produce oil.

Due to industrial activities, millions of hectares of land are barren and severely poisoned throughout the world. Less research has been done on the potential for grass to thrive in places where phytoremediation chemicals are present. Dicotyledonous herbs and shrubs with annual life cycles have been found to have the hyperaccumulation property or phytoextraction of metals. Since their biomass is mostly produced from the apical meristems, the biomass harvest of these species will thus be a single event with limited possibility for subsequent growth. Although it remains unaffected even after harvest, the regeneration potential is restricted to the base of the shoot above the root system. As a result of the perennial life cycles, continual phytoextraction of pollutants from polluted substrates can occur. This is advantageous to the phytoremediation process. Grasses frequently serve as the early successors in natural colonization in an ecosystem. The idea of planting edible plants on contaminated land should be outweighed by the potential for these plants to regenerate and their ability to adapt to a variety of environmental stresses. This is because the inclusion of the latter plants does not reduce the transmission of pollutants to the food chain through grazing livestock. After years of grass planting, the remedied area might be put to use for purposes like productive agriculture, or tourism destination, or it could be left alone to blend into the surrounding environment. The fraction of industrially contaminated areas is constantly rising, necessitating efficient phytoremediation techniques. When compared to a mixed culture, grass monocultures can occasionally be less effective at cleaning up a contaminated environment. Perennial grasses mixed with other crops can help degraded ecosystems regain their soil quality, biodiversity, and energy flow. In this regard, the development of a perennial plant cover in such locations requires the use of adequate genetic resource material.

7.3 EXOTIC GRASS

Certain exotic grasses can also function better on many degraded sites than native and site-specific grasses. Therefore, a supply of planting material for specific areas, which may include a variety of grass species, should be organized before a phytoremediation program. To use such genetic resources in remediation initiatives, efforts should be made to document them. Perennial grasses are among the most crucial plants for phytoremediation. This is because of their enormous biomass production potential, minimal fertilizer needs, and capacity to thrive even in soils with low fertility levels. Additionally, they offer a number of advantages like an improvement in soil quality and structure, a decrease in erosion, and an increase in biodiversity. Perennial grasses are far more sustainable to grow on damaged areas than conventional agricultural lands because the latter may encourage soil erosion and would not provide a sufficient economic return. Furthermore, from the perspective of an ecosystem, grasses constitute a group of ecologically dominating plant species since they are widely distributed on any waste or degraded area. The grass family is divided into aromatic and non-aromatic grasses. All of these grass species have been listed as prospective candidates for phytoremediation.

Perennial grasses have been documented by a number of authors to lower metal toxicity and stimulate phytoremediation. The majority of perennial grasses are terrestrial, herbaceous, or aquatic in nature, while others, like bamboo, are metal-tolerant plants. They store or phytostabilize metals in their roots, which gradually lowered the amount of metals in the soil. Some plants displayed Pb phytoextraction and grew successfully in Pb-polluted soil without reduction in biomass while Bermuda grass stabilized spill-affected soils. Bermuda grass exudates can combine with Pb to produce complexes. In addition to accumulating heavy metals, it has been observed that some grasses break down hydrocarbons. Grass species Axonopus can phytoremediate hydrocarbons from soil and may be successfully cultivated all over the world. When it comes to using grasses in phytoremediation, the grass species has an advantage over plants, shrubs, and trees due to its resilience to a wide range of climatic conditions (De Conti et al., 2019). Additional traits that make grasses suitable for phytoremediation include their drought-resistant nature, their presence of phytoliths, which are siliceous compounds found in epidermal cells and provide resistance too many abiotic stresses like metal toxicity, and their use as a feedstock for bioenergy and bioproducts. Since they can adapt to a wide range of climatic conditions, including extremely wet or dry environments, and have a larger application area, ornamental grasses have grown in popularity over the past few years. Ornamental grasses can be used for phytoremediation of polluted soils. They were tested in the field for years to determine how well they would survive in colder climates. Due to harsh environmental factors, restoration of degraded habitats frequently has its limitations. These plants can enhance the performance of neighboring target species. Grasses have a huge potential to be used as nurse plants in restoration management processes. For instance, the grass *Calamagrostis epigejos* demonstrated increased dominance and survival potential on bare ash deposits. In contrast to other arable crops, the grass Miscanthus can be utilized to cover the soil for a lengthier period of time since the organic matter inputs from the dropped leaves are anticipated to increase soil organic matter and improve soil structure. For the ecorestoration of a steel industry waste dump, we can cultivate *Cymbopogon citratus*.

7.4 AROMATIC GRASS

On the other hand, aromatic grasses are currently being advocated for phytoremediation due to their economic worth, flexibility, remediation capability, and minimal product contamination. *Cymbopogon citratus* can phytostabilize copper from mine when given supplements. *C. citratus* cultivated on a mixture of soil, sludge, and red soil could produce more oil, even while the oil's metal content was below the limits set by the Food Safety and Standards Authority of India despite the increased metal levels in the leaves. Tall Fescue intercropped with alfalfa removed soil polyaromatic hydrocarbons more effectively than alfalfa monoculture. Through phytomanagement, which is sought internationally, contaminated sites can be cleaned up and used for sustainable development. The demand for environmental remediation with a financial return is urgently in need of market prospects in phytoremediation. Aromatic grasses have been viewed as a crop that is extremely promising for phytoremediation initiatives. Because they are secure and their primary output, essential oil, is unpleasant to herbivores and free from the risk of metal toxicity. A variety of suitable aromatic grass species that are stress-tolerant and provide a large production of essential oil in such stressed areas are needed for profitable and effective phytoremediation. The key cause of climate change and global warming is the increase in greenhouse gas concentrations in the atmosphere (GHGs). After water vapor, carbon dioxide is the major GHG. The immediate advantage of producing biomass by cultivating perennial grasses in this situation is the storage of atmospheric carbon dioxide in the shoot and root. The grass is another benefit that can replace fossil fuels and is a significant source of feedstock for bioenergy. Additionally, these biomasses of perennial grass are a source of energy that might be considered to be renewable. Perennial grass biofuel crops are said to have higher carbon storage potential than agricultural crops and lower N_2O emission potential. Therefore, the ability of perennial grasses to mitigate climate change by producing bioenergy is the best option and can address multiple issues at once, including the reduction of CO_2 emissions, the mitigation of global warming, the reforestation of tropical areas, and the depletion of fossil fuels.

7.5 PERENNIAL GRASS

The potential for perennial grasses to conserve soil and prevent erosion is considerable. They also provide year-round ground cover. This is due to the fact that they have a vast root system that binds soil particles that, on sloppy fields, operate as porous vegetative barriers and a cheaper alternative to expensive soil physical constructions. By preventing soil erosion, vegetation barriers function similarly to physical soil structures. *Festuca rubra* has been widely utilized to stop erosion and stabilize slopes, hillsides, irrigation channels, and banks.

Figure 7.1. Phalaris arundinacea.

Source: *https://upload.wikimedia.org/wikipedia/commons/a/a4/Phalaris_arundinacea_1.jpg*

A cool-season perennial grass called *Phalaris arundinacea* (Figure 7.1) is also used to produce feed and stabilize soil. As an alternative, *Lasiurus sindicus* is a warm-season grass endemic to the Indian deserts that helps to stabilize blowing sand dunes and establish good rangeland. Because of their increased ability to stabilize soil surfaces, quick accumulation of soil organic matter, and effective storage and recycling of nutrients and ions, perennial grasses have been found to be more appropriate for phytoremediation. They exhibit a long grazing season, more feed, and a less combustible dry mass than other herbaceous plants. The biodiversity is increased by the aesthetic appeal. In deteriorated areas, they can also be utilized for horticulture reasons. Due to their extended growing season and lower initial fertilizer input requirements than annually growing plants, perennial grass can significantly reduce management in a developing ecosystem.

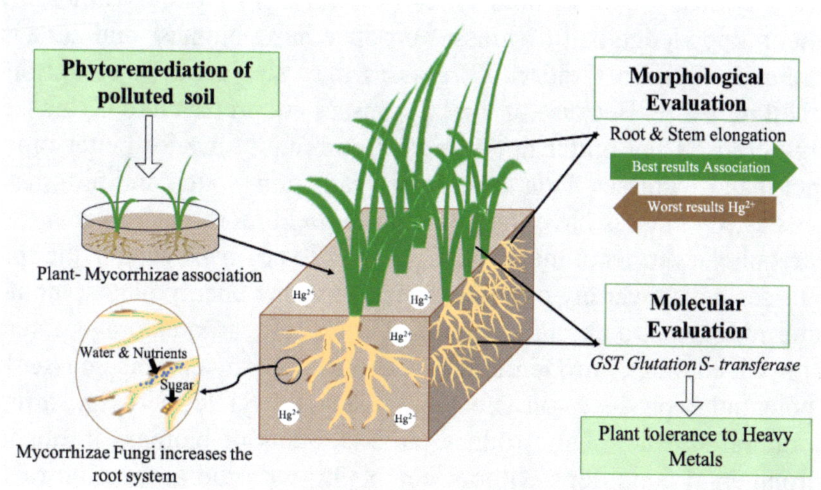

Figure 7.2. Mercury phytoremediation with Lolium perenne-mycorrhizae.

Source: *Leudo AM, Cruz Y, Montoya-Ruiz C, Delgado MdP, Saldarriaga JF. Mercury Phytoremediation with Lolium perenne-Mycorrhizae in Contaminated Soils. Sustainability. 2020; 12(9):3795. https://doi.org/10.3390/su12093795*

The process of phytoremediation of contaminated sites (Figure 7.2) involves cleaning up pollutants from the contaminated areas. However, in addition to this objective, the goals of such initiatives should also include the restoration of the native flora of the area. Utilizing local plants, particularly pioneer species that have already acclimated to these areas, effectively accomplishes this. The incorporation of such plants promotes the long-term succession of other species and aids in the development of a sustainable environment. This contrasts with the introduction of other commercial species, which frequently prevent native species from invading because of their competitive nature and frequently result in the development of weed. Native species that are already accustomed to the current nutritional conditions are able to remove contaminants while also flourishing with little care and displaying genetic flexibility to adapt to the environment. Many natural plants, though not metal hyperaccumulators, are metal tolerant; they uptake metals from the substrate at modest rates, detoxify them, and also generate a significant amount of biomass.

7.6 GRASS: THE SIGNIFICANCE

Grasses typically exhibit both types (C_3 and C_4) of photosynthesis. It is common knowledge that C_4 grasses produce more biomass and have more potential for carbon fixation. C_4 grasses may also grow in conditions of low soil moisture. Because of this, C_4 grasses are preferred to C_3 grasses as energy crops. They prefer not to have recurrent cutting for better biomass output than C_3 grasses. Due to increased radiation, water, and nitrogen-use efficiency, C_4 grasses have the potential to yield more than C_3 grasses, but they require a warmer climate than C_3 grasses to begin growing in the spring. C_4 grasses have a variety of advantages over other energy plants, including greater resistance to abiotic stressors, greater soil carbon storage potential, and improved water, nitrogen, and light energy use efficiency. C_4 grasses have the potential to produce consistent and sustainable yields across a variety of agricultural sites and can produce large amounts of biomass in uncertain environmental conditions without competing with the production of food crops. Many methods have been designed to enhance the physiological, genetic, and reproductive characteristics of grass species, potentially leading to the development of genetically modified grass species with high biomass production, high potential for remediation, and high tolerance to pathogens and environmental stresses. Numerous cultivars and hybrids with a high potential for biomass output have been developed by researchers working to improve grasses across the globe. The development of perennial grass cultivars with particular traits, such as high phytoremediation potential, high regeneration potential, the capability of quick regeneration after cutting, and resistance to stress factors, diseases, and pests, must also receive attention (Li et al., 2019).

Numerous possible perennial grasses have been used for a variety of purposes, including lawn maintenance, landscaping, livestock feed, golf courses, clean-up, providing plants on industrial waste dump sites, and restoring polluted and degraded areas. Ornamental grasses can also be utilized as potential genetic resources in the creation of parks on landfill sites. Perennial grasses have thus shown to be a valuable resource for securing livelihoods and reducing climate change. Native perennial grass species, on the other hand, have a high level of diversity, a capability for tolerating a variety of stress situations, a potential for regeneration, a high potential for biomass, and a potential for remediation in contaminated areas. The Industrial Revolution contributed to the socioeconomic advancement of the populace, but it also brought environmental degradation and contamination as a by-product. The majority of the world was covered in a forest at the time, thus the effects of industrialization on the ecosystem were not fully understood. However, as the population and industrialization grew, so did deforestation, and the need for more led to further environmental damage. There are now too many toxins in the soil, water, and air for both the present and future generations. The impacts of industrialization were recognized in the later decades of the 19th century when they were evident in a variety of human diseases, as well as livestock and plant diseases and toxicity. Since people have become aware of the negative impacts of pollution, many measures have been taken to lessen environmental contamination. For the purpose of removing pollutants, the procedures can be mechanical, chemical, physical, or plant-based. The least expensive and most environmentally friendly way for reducing contamination is plant-based remediation or phytoremediation.

7.7 REMEDIATION

Remediation using plants, bacteria, fungi, and their interactions with the surrounding environment is referred to as bioremediation. The manner of remediation using the various techniques varies depending on edaphic variables like soil, pH, EC, etc. as well as climate factors. As different plants and microorganisms develop in various seasons and climates, climatic variance indicates favorable conditions for plants and microbes to flourish. Because microorganisms make some contaminants available that plants would otherwise not have access to, the interaction of microbes and plants has been proven to improve the process of contamination remediation. Among the recognized plants, vetiver is better for its function in ecosystem phytomanagement, including preventing soil erosion, conserving water, phytoremediation, and also providing some economic benefits such as essential oil and the utilization of roots and shoots for a variety of uses. The World Bank originally used this amazing grass, which is native to India, to conserve soil and water in the middle of the 1980s. The grass has applications in the cosmetics, pharmaceutical, and fragrance sectors. Due to its pharmacological, phytoremedial, and phyto economic capabilities, the vetiver plant is our primary focus. Biologically active chemicals derived from plants are also required for the scientific confirmation of their traditional therapeutic and phytoremedial efficacy. The fact that vetiver may grow on land, in wetland systems, and in aquatic systems strengthens its significance. This plant has demonstrated its effectiveness in all situations by effectively removing pollutants.

The plant has demonstrated its use in agriculture in addition to its ability to remove pollutants. Vetiver lowers the concentration of nutrients like N, P, and K in soil and water. Vegetable roots naturally have the capacity to form symbiotic relationships with the rhizosphere microorganisms. Through the oil in its roots, vetiver is also known to decrease termite populations. Although vetiver appears to be a small herb, its significant contribution to phytoremediation is undisputed. Vetiveria is significant not only for its essential oil and potential for phytoremediation, but also for its function in soil and water conservation. With its deep roots, this plant has been shown to lessen soil erosion. Additionally, this increases ground water recharge and enhances soil water retention. Vetiver can be cultivated in homes for waste water treatment because it doesn't need special care to grow. As a result, less contamination will spread to surrounding places since the toxins in residential wastewater will be reduced. The root and shoot biomass of vetiver is strong. The roots of vetiver are used to make essential oil, which is utilized in perfumes, medicine, and cosmetics. The residual portions of the root are not waste; rather, they can be used to produce energy and biogas.

7.8 VETIVERIA FOR PHYTOREMEDIATION

Vetiveria shoots have a healthy amount of biomass and are rich in lignin, cellulose, and hemicelluloses. The processing of the shoots will produce energy, such as bioethanol. It has been observed that vetiver shoots can produce biogas. The method of phytoremediation involves using plants to draw toxins out of the soil, water, and air. The method is economical and safe for the environment. The perfect plant, vetiver, can do a variety of tasks. The plant vetiver is capable of rhizoremediation, phytoextraction, phytovolatilization, phytostabilization, and other types of remediation. Vetiver is a plant that has been specifically chosen for a variety of uses. Any soil type, including sodic, alkaline, acidic, coal, and gold mines, can support the growth of vetiver. These plants now have significance due to their capacity to accumulate pollutants. The essential oil that vetiver produces can absorb heavy metals and other impurities making it a wonderful gift from nature that plays a crucial part in phytoremediation. Heavy metal-contaminated soil, such as that with Pb, Zn, Cu, Fe, and Mn, can support the growth of vetiver. Both roots and shoots have been reported to accumulate them, with roots having a higher percentage than shoots. High amounts of radioactive compounds in soil were caused by the growing usage of radioactive materials in weapons, power, and fertilizers. Because of the substantial metal burden already present in the soil, this leads to a dangerous situation. Soil remediation is essential right away. Vetiver grows well in soils with uranium and does not exhibit any harmful symptoms even at the greatest levels of contamination in regard to the remediation of radioactive chemicals. The biomass of vetiver was unaffected by the presence of uranium, and higher accumulation was found in the root than in the shoot.

Vetiver may remediate phenol at 200 ppm in just 4 days. Given that the experiment was carried out in an aseptic environment, vetiver was the only agent involved in the elimination of phenol (Jisha, Bauddh, and Shukla, 2017). In an experiment with vetiver on a floating platform and in a wetland with horizontal flow, it was discovered that the decomposition of phenol was slower on the floating platform than it was on the horizontal flow. Ethidium bromide (EtBr), a significant laboratory by-product that is used to stain DNA, is poisonous and mutagenic. This substance is used in laboratories all over the world, and it needs to be properly disposed of because, if discharged, it could harm the local species. A similar hazardous substance is acrylamide used to make polyacrylamide gels for RNA and DNA analysis. If acrylamide enters soil or water, it is a neurotoxic and carcinogen, which is damaging to flora and fauna. Thus, it is crucial to safely remove these chemicals, and it has been demonstrated that vetiver grass can do so. Conjugated atrazine was detected in vetiver leaves, while alkylated atrazine was discovered in the root and shoot. Atrazine has been seen to accumulate and get sequestered in the lipid portion of plants. As a result, vetiver oil has a concentration of atrazine. Microbes have a significant role in the degradation of atrazine in conjunction with the roots of vetiver.

7.9 ANTIBIOTICS

One of the significant antibiotics that are increasingly used in both laboratories and medicine is tetracycline. The unintentional discharge of this contaminates soil and water and can occasionally increase microbial community tolerance or reduce microbial diversity. As a result, it needs to be remedied, and the first step should be removal. Tetracycline is eliminated from the hydroponic solution by vetiver. One of the main micro-pollutants is naproxen and it is a medication that is used to alleviate pain and inflammation.

Due to their efficient power output, nuclear power plants are currently a solution for the electricity issues in both emerging and developed nations. But this also results in radioactive wastes that are bad for the surrounding flora and fauna. There is a need to remove the radioactive wastes from the environment since they have long-lasting effects across large areas. It was discovered that vetiver worked well to remove strontium (^{90}Sr) and cesium (^{137}Cs) from a solution that had been contaminated with these radioactive elements. Vegetables can restore up to 95% of the Cs-contaminated soil. Vetiver grass can be a useful plant in the clean-up of nuclear waste. In contaminated sites, the soil close to the oil drilling site is heavily contaminated with oil. On extremely polluted oil sites, vetiver is discovered to flourish and it can break down the hydrocarbons in oil (Sharma, Singh, and Manchanda, 2015).

The combustion of coal produces fly ash. Human existence now depends on electricity, which has led to the construction of numerous power plants, the majority of which use coal as their primary fuel source. There are several by-products produced when coal is burned, but fly ash accounts for 15–30% of them. While some fly ash is utilized in the production of cement, bricks, roads, and land reclamation projects, it is also employed as an agricultural supplement. The majority of the residual fly ash is dumped in open landfills, which contaminates the soil and water. This explains why these landfills are occasionally referred to as fly ash ponds. There is a significant risk to the way of life because fly ash contains a number of contaminants and heavy metals that seep into the groundwater and surrounding water sources. As a result, cleaning up these fly ash ponds is urgently necessary to preserve the livelihood from the negative consequences of contamination. Both mechanical and plant-based systems are available for the removal of impurities; the former is expensive while the latter is less expensive and environmentally benign. The screening and identification of plants that can grow effectively in fly ash deposits and heavy metal-contaminated areas, as well as accumulate them with efficiency and speed, will be necessary for plant-based techniques. The fly ash ponds have been found to support vetiver grass growth. Therefore, vetiver was chosen as a key plant in the clean-up of fly ash ponds.

7.10 CONTAMINANTS

The high levels of contaminants in soil and water are having a negative impact, as seen by the extinction of species and loss of species variety. Eutrophication, or the death of the aquatic ecosystem, is brought on by the introduction of contaminants into the aquatic system. A plant-based approach works well and is affordable, and vetiver grass helps. In comparison to other plants, vetiver grass has been shown to take nitrogen and phosphorus from the water more quickly. A combination of different plants and microbes that absorb atmospheric carbon will have a significant impact on halting global warming. Climate, soil characteristics, and regional land use are only a few of the variables that both support and counteract the process of carbon sequestration. Carbon is known to be sequestered by trees, lemon grass, and palmasora, but vetiver sequesters carbon at a high rate. One of the best appropriate plants is vetiver, which has significant economic value in addition to being suitable for easy growth. Thus, vetiver has proven to be the most effective in addressing climate change and the global warming problem due to its multipurpose application and multidirectional approach. Vetiveria has an extremely high capacity to sequester carbon when compared to other trees. It will not only help to sequester carbon but also bring in money for the growers. Vetiver oil is in extremely high demand. In rural places, the residual shoot and root are utilized to thatch dwellings. The risk of herbivory is also reduced because this plant is infrequently grazed. Vetiver thrives in areas with poor soil fertility, such as wasteland and damaged lands (Brandt et al, 2006).

The grass known as vetiver has been utilized to conserve soil and water. Intercropping with other plants enhanced soil structure, decreased erosion, and improved soil water retention. In arsenic-contaminated soil, intercropping vetiver with crop plants also helped crop plants absorb less arsenic. The vetiver root and shoots contained an arsenic buildup. Vetiver arecanut system boosted the productivity of arecanut (*Areca catechu L.*). Vetiveria plays a variety of responsibilities, from environmental clean-up to insect control. With the help of its powerful essential oil, vetiver has been utilized to successfully eradicate termites. A substance called nootkatone, which is found in the roots of vetiver, has the power to affect termites' behavior. This is due to the nootkatone's ability to kill protozoa in termite guts. Using vetiver oil causes termites to build fewer and shorter tunnels. Higher cocoa plant growth and used vetiver grass compost can reduce termite populations. Vegetables with vetiver mulch have a lower fungus attack rate. The most crucial elements of an ecosystem are water and soil. They are disturbed, and the ecosystem's vegetation and wildlife are impacted. On slopes or bare terrain, soil erosion is typical following heavy rains. Erosion is lower on soil that has been planted, but as the population and demands on the ecosystem have grown, deforestation and contamination have increased, and bare lands are now more prevalent, erosion has increased. Reduced variety, irregular rainfall, global warming, the loss of fertile soil, and decreased groundwater recharge are all signs that the ecosystem's equilibrium is being upset. Although the issue can be resolved by planting trees, erosion can only be prevented in slope areas, earthquake, and hurricane-prone regions, and by planting plants that can endure these effects.

There have been reports of increased water conservation and soil nutrient levels in vetiver plantation areas. Vetiver plantations reduce soil runoff by roughly 69 percent when compared to areas without them. The efficiency of nitrogen usage was also improved. Some nations (China, Malaysia, Thailand, Venezuela, and Argentina) grow vetiver to stabilize the side of the road (Dudai et al., 2018). Vetiver is planted in coastal areas, slopes, for road side stabilization, near ponds to decrease soil erosion and runoff, and for water conservation in countries like India, Thailand, Vietnam, Madagascar, South Africa, and Uganda, among others. Despite having so many benefits, it does have some drawbacks. Vegetables like vetiver have broad, underground roots. When the roots are removed, the soil becomes more susceptible to flow and erosion. In order to prevent unchecked cultivation by the organizations participating in the vetiver trade, sufficient vigilance must be maintained. Extreme vigilance should be exercised when disposing of the biomass for phytoremediation reasons. As there have been reports of reduced contamination levels in the oil, it must also be checked for vetiver oil that was produced from polluted areas. Vetiver has a deep root system, which has led to the notion that it can compete with surrounding plants for nutrients and moisture and stunt their growth.

Vetiveria has deep roots and develops microbial associations with the rhizosphere microbes. This enables vetiver to flourish in soils with very low nitrogen concentrations. Vetiver has an abundant root system, bushy shoot growth, and high biomass. Since vetiver is a C_4 plant, it efficiently uses CO_2 through the Hatch Slack pathway at high light intensities. It also has a quick rate of growth, producing more biomass faster. The majority of plants have a serious problem with infection to diseases and pests. Vetiver is resistant to a number of diseases and pests. Leaf blight has been reported by the USDA, along with several fungi. Vetiveria is sterile and non-invasive; it blooms but does not produce seeds since it is sterile. As a result, it cannot spread to the native species because no seeds are produced, which limits its ability to do so. Additionally, the roots grow downward in a straight line rather than outward. Vetiver can live up to 150 years and has a lengthy lifespan. Regarding phytoremediation, this is a crucial prerequisite as it will aid in the long-term elimination of toxins at heavily contaminated locations. The plant also grows with a lot less maintenance. Vetiver has a deep root system and lengthy roots. It helps to prevent soil erosion. The roots can grow up to a length of 3–7 meters.

This large root network also aids in soil anchoring, which prevents soil erosion. This promotes water efficiency. Mycorrhizal partnerships can occasionally form at the roots, which additionally aid the plant in improved water and nutrient uptake. Plants with both xerophytic and hydrophytic traits can thrive in a variety of environments, expanding their growing territories. It has a lot of biomasses underneath as well as above ground. The roots extend widely and penetrate deeply. The main characteristic of vetiver is that it is a C_4 plant, which means it will lose less water and use carbon more effectively. Due to their pleasant and nutritional value, grasses serve as a significant source of feed in arid settings. Rangelands are significant habitats for grasses, particularly perennial grasses. In India, Pakistan, and the Arab world, all types of desert perennial grasses are extensively used as animal fodder. However, L. sindicus is given more attention in the Indian state of Rajasthan where it serves as a source of cover for the soil and forage for cattle. There is more possibility for producing fodder sustainably and cheaply using natural forage plants. In addition to being used as fodder, *L. sindicus* can be utilized to reduce soil erosion and improve rangeland productivity by restoring or rehabilitating degraded desert grasslands.

Therefore, L. sindicus is one of the well-known and most promising grass species for forage production in hot and dry desert environments. It thrives well in arid climates and yields a lot of fodder. It is one of the earliest grass species available for livestock and is one of the most xeric in nature, drought tolerant, and nutrient-rich desert grasses. Its green patches that appear on sand dunes and interdunal sandy areas after rainfall demonstrate its superior ecophysiological capacity to other parent grasses. Choosing natural grass species to stabilize hot, arid desert areas is a sustainable way to manage and recover a region in order to increase grass productivity. The perennial grass *L. sindicus* is a species of the Indian desert and is considered to be one of the most prevalent and natural inhabitants of the hot, arid habitat of the Thar Desert. It is highly nutritious and well-tolerated by livestock in the desert.

It naturally grows in a variety of dry regions throughout Asia and Africa. Its height is from 30 to 60 cm, but under ideal circumstances, it can reach 90 to 100 cm. This crucial grass for agriculture can withstand droughts, but it is in grave danger of going extinct because of shifting land use patterns, an increase in soil moisture regime, and overgrazing. Due to its excellent nutritional content, livestock in the desert region specifically eat it. It aids in the stabilization of sand dunes and the growth of healthy grasslands. This grass' ability to survive in the hot, arid ecosystem of the Thar Desert is severely threatened by severely overgrazed pasture, land-use change, and increased aridity. Rhizospheric microbial investigations of desert plants are less well documented, despite the fact that the rhizospheric influence is qualitatively and quantitatively more pronounced in desert soils than in soils of wet climates. Due to their ability to withstand extremes in temperature and rainfall as well as deteriorated soils with low organic content and few supplies of accessible inorganic fertilizers, desert plants are able to survive.

7.11 DESERT PLANTS

Desert plant rhizosphere microbes need to maintain adaptations to cope with challenging environments such as recurrent droughts, desiccation, high temperatures, and high osmolarity. An important task is to investigate in situ biodiversity conservation linked to such niches to support ecological processes in hot and dry desert ecosystems through the study of the rhizosphere of Sewan grass. Despite having a tolerable nature, it needs protection while it is still establishing itself. Sewan grass can be found in areas that have been exposed to heavy metals. On the sandy plains, low dunes, and hills of this area, it thrives under moisture and temperature stress. Animal husbandry continued to be the primary occupation of the locals in the Indian desert's Sewan grasslands, which sustained an ever-increasing livestock population. There is a need to examine how Sewan grass regenerates in relation to rhizosphere soil moisture levels. It is a perennial C_4 grass that grows to a height of about 1.2 m and is upright and branched. In xeric settings, amphistomatic leaves are more prevalent and have better photosynthetic rates than hypostomatic leaves. The linear, 20–45 cm long, silver–green–yellow leaves have a setaceous tip. A slender leaf blade alternates with the leaves. The anatomy of the leaves is characterized by the C_4 NADP-ME type, and they have grown sclerenchyma to provide mechanical strength in drought and wind-prone conditions. The inflorescence is a raceme that is 10 cm long, silky white and bears 3 hairy spikelets at each node. The biomass of this grass is increased by all of these physical traits. *L. scindicus* can live up to 15 years and is xeric in nature and resistant to drought. When drought and winter are dominant, Sewan grass exhibits a high root-to-shoot ratio, demonstrating that its roots are less sensitive to these harsh conditions (Radwan, 2009).

L. hirsutus is mostly propagated by sowing and is extremely drought-tolerant. This thick-rooted, desert perennial grass prefers non-alkaline sandy soil and can grow up to 1 m tall under ideal conditions. To survive the harsh climatic conditions, this grass has evolved a variety of morphological, anatomical, and biochemical adaptations. Because Sewan grass is native to the area and has strong potential for adaptability to desert conditions, it should be planted more frequently in hot, dry landscape designs. It is a very valued and excellent grass for landscape design in hot, arid regions due to all of these characteristics. In order to effectively remove or stabilize metals, phytoremediation is a viable option that must focus on three key objectives: minimal input, economic yield, and least risk. Due to their potential for stability, grasses, which are members of the Poaceae family, have recently gained increasing attention in relation to phytoremediation. Because lead is poorly transported from roots to shoots through the vascular system, it accumulates more lead in roots than leaves, making it a suitable candidate for phytostabilization programs. Because roots accumulate more lead than leaves do, *L. scindicus* can be employed in Pb phytostabilization. Pb was found to have negative impacts on the development and metabolism of *L. scindicus*. Mycorrhizal colonization, nutritional content, and heavy metal concentration of naturally grown grass *L. scindicus* on gypsum mine debris were all reported. On mine debris compared to regular soil, *L. scindicus* grass had a much higher percentage of arbuscular mycorrhizal fungus (AMF)-inoculated roots. In comparison to typical soil, *L. scindicus* grass on mined debris had significantly greater K, Ca, Mg, and Na nutrient concentrations and lower N and P concentrations. Due to their low availability in mine debris, N and P are present in lower concentrations. The results demonstrated the viability of employing AMF for the rehabilitation of mining waste. AMFs are widely known to aid in the mobilization and uptake of nutrients by plants.

7.12 SEWAN GRASS

Sewan grass has been identified as an endangered desert grass species because it is the most valuable grass in the Thar Desert region of India.

Sewan grass faces a tremendous threat to its survival in the Indian Thar ecosystem due to severe overgrazing, agricultural modernization, alternative land-use patterns, frequent droughts, and increased human activity. Therefore, it is crucial to preserve its biodiversity in order to fulfil future goals in light of shifting environmental conditions (Prasad and Tewari, 2016). Morphological, biochemical, and molecular techniques are among those employed to identify genetic variability in vulnerable species. In order to preserve diverse germplasm for future demands, the genetic diversity of Sewan grass has recently been investigated using the molecular markers technique. Less harsh conditions and the availability of adequate irrigation were found to be the cause of the higher diversity based on allele content and its frequency, whereas adaptive pressures may be a limiting factor. Information on the rhizosphere microbiology of Sewan grass is quite limited, despite the fact that the rhizosphere effect is qualitatively and quantitatively more apparent in hot, arid desert soils. The rhizospheric microbes associated with desert grasses should have the ability to adapt to certain stresses, such as frequent droughts, high temperatures, starvation, high osmolarity, and desiccation, as desert grasses must endure extremes in temperature as well as grow in typically poor soils with low organic content and few bioavailable inorganic nutrients. Numerous studies have focused on the variety of nifH sequences to examine the significance and function of nitrogen fixation in ecologically distinct terrestrial settings. Lasiurus roots were investigated for possible associations with Azotobacter and Azospirillum. It has been demonstrated that the high productivity of Sewan grass in dry and semi-arid regions is due to healthy populations of free-living, nitrogen-fixing bacteria like Azotobacter and Azospirillum. Rhizospheric microbiology is important for biodiversity preservation and niches that can withstand mild ecological processes in the oligotrophic desert ecosystem.

For the growth of self-sustaining plant cover on degraded sites or garbage dumps, perennial grasses are suitable candidates for eco-restoration and phytoremediation. Due to its remarkable ecological adaptability, Sewan grass (*L. sindicus*) is a very important and ideal perennial pasture grass in hot, dry areas. It is widely utilized as feed grass in pastures, particularly in extremely arid desert locations. Understanding its remarkable potential for survival and ecophysiological adaptability to harsh environments requires a lot of work. However, it is necessary to investigate the methods by which the roots and shoots of L. sindicus growing in polluted areas accumulate, distribute, translocate, and sequester inorganic and organic contaminants. The ability of this grass in a contaminated region to encourage microbial activity through root exudates must also be determined. Understanding the relationship between new genes and L. sindicus's adaptive response to contaminants can be improved through genetic investigations. Therefore, choosing the best phytoremediation techniques and increasing biomass production can be better understood by using physiology, biochemistry, molecular biology, and proteomics in conjunction with genetic engineering and breeding of *L. sindicus*. Utilizing *L. sindicus* in phytomanagement practices provides excellent opportunities for formulating policy frameworks that support the remediation of polluted sites as well as biomass production in hot, dry, arid regions around the world.

7.13 POPULATION: THE WARNING

Population growth and pollution levels are inversely correlated with the planet's carrying capacity and ecosystem health. Rapid industrialization has led to pollution. The natural biogeochemical cycles are being impacted by the toxins that are being added to the soil and water. Heavy metal contamination is a severe issue because they do not dissolve and can convert into various forms, making the substance more dangerous. Therefore, it is important to lessen and clean up these heavy metals. These contaminants can be reduced using mechanical, chemical, and plant-based methods, or phytoremediation. The preceding methods outlined have advantages and drawbacks. Mechanical and chemical methods of removal are expensive and do not offer a long-term solution for the issue. Even though the impurities from these two procedures can be eliminated quickly, they are expensive. Although phytoremediation is an economical and environmentally benign method of removing toxins, the remediation process takes a lot of time. The utilization of plants that produce energy (bioethanol, biogas, or biofuel) or commercial goods (essential oil, etc.) in addition to phytoremediation can overcome the long-time limitation.

It is time to reconsider alternative energy sources in order to tackle the current situation of climate change and the varied repercussions of pollution on the flora and fauna. Additionally, the global need for energy is growing. The current energy and climate change problems will be resolved by the usage of renewable energy sources. The usage of renewable energy sources like solar and wind is widespread, but not enough in all nations because these require expensive setups that developing nations cannot wage. Since every country seeks to do more with fewer resources, if a cheap source of energy production is made available, both rich and developing nations will be able to afford it. Plants that produce renewable energy are in question here because they have good biomass, which can be used to make fuel. It would be advantageous since both issues would be resolved if the plant utilized was also an effective pollutant accumulator. Miscanthus species are outstanding plants that have the capacity to produce both phytoremediation and bioenergy. In Europe, miscanthus species are frequently utilized to produce bioenergy. Because Miscanthus has good biomass, it also satisfies one of the fundamental criteria for plants used in phytoremediation. Miscanthus is the plant for phytoremediation since it is perennial, herbaceous, grows quickly, and has strong root and shoot biomass.

The information that is now available from some of the nations that have begun using plant biomass for the generation of electricity and fuel will help make this more evident. There is an urgent need to boost energy production to meet the demand, which is feasible through alternative expensive means of energy production but may have an influence on the climate. Utilizing plant biomass for energy production is a solid strategy right now to produce energy with the least amount of environmental impact. Miscanthus species are employed for both biomass-based bioenergy and phytoremediation.

CHAPTER 8

Adaptation of Plants

CONTENTS

8.1 INTRODUCTION

The industrial and agricultural revolutions in recent decades have increased the levels of toxins in our environment, which are now key factors in plant and animal toxicity. Among other pollutants, rising concentrations of salts, heavy metals, pesticides, and other chemicals represent a hazard to agricultural and natural ecosystems. These toxins cause pollution of the soil, air, and water as well as a reduction in the amount of arable land. By altering the structure of natural communities and ecosystems, they also contribute to changes in species composition and biodiversity loss. Different strategies are being used in this situation to restore degraded ecosystems. One of these has the ability to eliminate harmful toxins from the environment and it is called phytoremediation. This strategy is based on the usage of naturally occurring hyperaccumulator plant species, which can withstand relatively high amounts of environmental contaminants. It is possible to harvest and remove from the site the pollutants that have collected in the stems and leaves of high-tolerant plants. Therefore, by harvesting the biomass, this method can remove significant levels of pollutants. However, with a better understanding of the physiological, biochemical, molecular, and genetic techniques of plant resistance to both naturally occurring and anthropogenic toxins, the efficacy of phytoremediation approaches can be boosted. Environmental degradation-related problems are getting worse as the population grows. Most pollutants are difficult to breakdown; therefore, they build up in the environment. Degradation of the soil, deforestation, desertification, loss of species diversity, pollution, acid rain, and the greenhouse effect, are caused by the accelerated accumulation of toxins (Barocsi et al., 2003).

8.2 TOXINS

Inorganic and organic substances known as toxins are harmful to plant growth and metabolism. Because of human activity, they are released into the environment. For instance, during the extraction process by mining, smelting, combustion, and industrial effluents, salts and heavy metals are released via leaks. Soil pollution has been caused by the extensive use of pesticides and fertilizers in agriculture. Other pollutants, including those made of petroleum, explosives, cyanides, etc., are likewise very harmful to living things. Pollutant toxicity is assessed in relation to its biological function, capacity for bioaccumulation, and persistence in the environment. But each of these factors varies significantly for various pollutants according to their molecular composition and physical and chemical characteristics.

This is in contrast to organic pollutants, which are eventually transformed into CO_2 and H_2O. As a result, their removal is far more challenging than the removal of organic contaminants and calls for a different approach. Metals can also move freely across different environmental compartments. These characteristics render them highly harmful to the environment and human health if they are present in the environment in larger amounts.

The ability to absorb, collect, or detoxify different organic and inorganic contaminants can be found in a variety of hyperaccumulator species from different groups of bacteria, fungi, lichens, and higher plants. Toxic materials are removed from various environmental components, especially from soil and water, using this process, which is generally known as bioremediation. All of these tactics are efficient for removing a particular pollutant and are based on various methodologies. Additionally, a certain approach that gets rid of one pollutant well could be completely ineffective in getting rid of another. For the elimination of salts and heavy metals, for instance, phytoextraction and phytoimmobilization may be effective. However, phytotransformation or phytodegradation may be more efficient in the removal of organic pollutants such as hydrocarbons and explosives. As a result, the choice of a specific plant species cultivation in contaminated areas depends on the type of contamination, the mechanism utilized by that species to remove the contaminant, the plant tolerance to the pollutant, and other environmental factors.

8.3 POLLUTANTS

Biodegradable (organic) and non-biodegradable (inorganic) pollutants are the two main categories of toxins. When living organisms participate in the biogeochemical cycles, biodegradable poisons are quickly converted into simpler molecules by the activity of these organisms. Because they are present in our environment in such small amounts, these poisons are typically not dangerous. However, they can be extremely harmful to all living things in large doses. Organic poisons like petroleum compounds are poisonous even in little amounts. Domestic and agricultural waste, petroleum products, urine and feces, and sewage water are examples of biodegradable pollutants. Living organisms cannot break down non-biodegradable poisons into basic, harmless compounds. These substances include salts (NaCl), oxides of nitrogen and sulfur (NO_2 and SO_2), cyanides, and inorganic fertilizers, pesticides and insecticides (DDT), heavy metals (nickel, mercury, copper, lead, aluminum, arsenic, etc.). These toxins can stay in the environment for a very long time and end up harming organisms when they enter the food chain. As a result, the removal of these toxins from the environment is much more difficult than that of biodegradable ones (Glick, 2020). The immediate environmental component that is subject to the deterioration by these poisons is reflected in this classification system. Salts, heavy metals, inorganic and organic fertilizers, pesticides, residential waste, agricultural waste, and industrial pollution are toxins that build up in the soil. Similarly, main (CO_2, CO, SO_2, NO_2, CH_4, ammonia, volatile organic compounds) and secondary (ozone, peroxyacetylene nitrate) air pollutants are toxins that are discharged into the air. Sewage water, leftovers from food processing plants, industrial waste, petroleum products, fertilizers and pesticides from agricultural runoff, etc. are the main causes of water pollution. The majority of the toxins can easily move from one component of the environment to another before accumulating in soil and water bodies. These contaminants can then be readily absorbed by aquatic life and plants, where they enter the human body and cause severe infections and diseases.

The main environmental contaminant contributing to the global salinity issue is the high salt content of various soil profiles. High salt concentrations are thought to have an impact on 20% of the world's arable land and around 7% of the planet's overall land area (Hopmans et al., 2021). A significant portion of the irrigation system is also severely salinized and unsuitable for the production of agricultural crops. NaCl and $MgSO_4$ are the two salts that cause soil salinity issues frequently. Soil salinity can be divided into two categories: primary or natural and secondary or induced soil salinity. Primary or natural soil salinity results from the weathering of minerals obtained from parent rocks that are extremely salinous. Secondary salinization happens when people disturb the water cycle. It happens when seasonal crops with shallow roots take the place of native perennial plants. Additional practices like excessive grazing and deforestation also limit plant cover. This causes the subsurface water table to rise by up to 2-3 meters, and capillarity causes the salts that have been dissolved in various soil profiles to rise to the surface, leading to secondary soil salinity. With the decrease in vegetative cover, more water enters underground aquifers, while significantly less water is absorbed by plants. Agronomic methods that are widespread and the introduction of alien crops can sometimes change how much water the plants need. If this causes subterranean aquifers to recharge more quickly than they discharge, the groundwater level will rise and carry salts with it, leading to secondary soil salinity. Although a high salt content in the soil can affect crop plants in a number of ways at the biochemical, molecular, and physiological levels, the most frequent effects on crops include photosynthesis inhibition, nutrient imbalance, changes in metabolic activities, disturbances in solute accumulation, enzyme activities, hormonal imbalance, and others. Salinity inhibits plant growth in different ways: salt-induced water stress; specific ion toxicity; oxidative stress, i.e., production of reactive oxygen species; and hormonal imbalances. The length of the stress, the stage of the plant's growth, and the species of the plant all affect how much growth is inhibited by salt stress. But compared to later adult stages, early growth stages like germination and seedling stages are thought to be more vulnerable to salt damage.

8.4 PLANT RESPONSE AND SALT STRESS

One of the most notable plant responses to salt stress is the accumulation of compatible solutes, which are organic compounds with low molecular weight. This phenomenon aids the plant in adapting to various stressful environments. Many suitable osmolytes, including proline and glycine-betaine, are found in a variety of plants and are thought to be quite effective at controlling growth in stressful conditions. Even though they build up at high cellular concentrations, these suitable solutes are non-toxic, have a low molecular weight, and are highly soluble. They shield cellular structures from damage brought on by abiotic stress. For instance, in plants exposed to stressful conditions, they support osmotic adjustment, scavenge reactive oxygen species, stabilize enzymes and proteins, and safeguard membrane integrity. In recent years, heavy metals have drawn a lot of attention as a potential environmental hazard. This is a result of their excessive use in various industrial processes, and as a result, their toxicity is more prevalent. The majority of metals are frequently employed in a wide range of industrial operations, including the production of steel, batteries, alloys, electroplated metal parts, insecticides, and textile colors.

8.5 HEAVY METALS

Of the 35 metals that pose a threat to environmental health, 23 are referred to as heavy metals. They are iron (Fe), gold (Au), lead (Pb), nickel (Ni), manganese (Mn), mercury (Hg), platinum (Pt), silver (Ag), thallium (Tl), tellurium (Te), bismuth (Bi), cadmium (Cd), cerium (Ce), copper (Cu), chromium (Cr), cobalt (Co), gallium (Ga), antimony (Sb), bismuth. Among these, arsenic, lead, mercury, cadmium, nickel, iron, and aluminum are the most prevalent heavy metals that are hazardous to both plants and animals.

The majority of metals are quickly absorbed by plants and bioaccumulated in various organs. Through the consumption of food, drinking water contaminated with metal, or breathing in air containing harmful metals, these metals may eventually find their way into the human body. Since some metals function as micronutrients in small quantities and are therefore regarded as important nutrients, not all metals are hazardous. Due to their extremely low concentrations or requirements in biological systems, several metals such as iron, copper, manganese, and zinc are sometimes known as trace elements. They are present in sufficient proportions in our foods, fruits, and vegetables since they are naturally present in the soil. They are a part of widely accessible multivitamin products. The majority of metals serve as cofactors in a variety of metabolic processes. For instance, among the common metals with recognized biological activities in plants are Fe, Zn, Cu, Ni, and Mo. Most of these metals are needed as enzyme activators, and some of them are even essential parts of a variety of metalloenzymes. Because of this, their deficiency may prevent plants from growing and developing normally, which manifests as chlorosis and subsequent necrosis of plant tissues.

Some biologically non-essential metals can be hazardous to organisms even at very low doses. The general symptoms of metal toxicity in plants include stunted shoot and root growth, poor branching system, deformation of various plant parts, abnormal flower shape, decreased biomass production, leaf spots, root tip anomalies, inhibition of germination, and chlorosis that can ultimately result in a decrease in the yield of agricultural products. In order to protect crops from pests and pathogens, herbicides and pesticides have long been employed as the most efficient methods. They include algicides, fungicides, bactericides, insecticides, weedicides, and herbicides. To protect seeds, grains, and cereals from pests and diseases and to increase their storage capacity, these chemicals are administered to crops at various growth stages, such as before seeding, during crop cultivation, and after harvest. These substances are used as sprays, dusts, powders, aerosols, seed treatments, and oil-based solutions. DDT, benzene hexachloride, lindane, malathion, 2,4-dichloro phenoxy acetic acid, etc. are a few examples. The majority of pesticides are successful at controlling pests and pathogens, making them the most widely used, cost-efficient, and efficient crop protection method worldwide. Although using these chemical compounds to control pests and pathogens is thought to be an efficient method, doing so can have negative impacts on plants and animals, including invertebrate and vertebrate species. During their preparation and processing steps, application techniques, post-application evaporation and volatilization, and water runoff, these pesticides and herbicides can enter the atmosphere and ecosystems. A significant source of the environmental contamination caused by outdated chemicals is their disposal in soil and water bodies. Insecticides are the class of chemicals that have the greatest impact on the environment and the toxicity of living things.

8.6 EFFECTS

Several characteristics affect how poisonous or destructive pesticides and herbicides are to species and the environment. These include their chemical composition (systemic or non-systemic), the active components, exposure, their persistence in the environment, and their concentration.

In addition to these facts, other elements like the farmer's expertise as the applicator, the time of application, and the weather condition have a substantial impact on the actual toxicity and can make it exceedingly dangerous. These substances build up in soil and water bodies and are exceedingly harmful to both people and non-target organisms including plants and animals. Organic substances known as cyanides have the cyano group (C-N) in their structure.

Prussic acid poisoning is another name for cyanide toxicity. Hydrogen cyanide (HCN), potassium cyanide (KCN), and sodium cyanide are a few different types of cyanides (NaCN) (Burken, 2003). Among these, HCN is an odorless gas with a bitter-almond-like scent, while NaCN and KCN are white powders with a comparable odor. When NaCN and KCN are combined with water, they both break down into HCN, which is harmful to living things. Certain bacteria, fungi, algae, and higher plants contain naturally occurring cyanogenic chemicals. Many plant groups, including Poaceae, Papilionaceae, Sambucaceae, Euphorbiaceae, and Rosaceae, naturally contain cyanogenic chemicals.

They can be found in trace levels in a variety of plant fruits, including bitter almonds, mango, peach, plums, apple seeds, and citrus seeds. Cyanides play a significant part in the defence of plants against herbivory and are typically found in bound forms in plants called cyanogenic glycosides. For instance, it has been shown that cassava roots contain an excessive amount of cyanogenic glycosides.

Amygdalin is the best-characterized cyanogenic glycoside among those discovered in plants. It is found in a variety of plant species, particularly in the leaves and seeds of cherries, almonds, and peaches, among others. More than 3000 plant species have been documented to have cyanogenic glycosides, which means that ingesting these species by people or animals could result in HCN toxicity.

8.7 SOURCES OF CYANIDES

Various industrial sources also release cyanides, in addition to natural sources. For instance, cyanogen halides are released when water containing free cyanides is chlorinated or bromated, whereas thiocyanate is released in a variety of industrial effluent discharges. In addition to chemicals such as nylon and plastic, adhesives, cosmetics, dyes, computer electronics, pharmaceuticals, pesticides, rodenticides, wine, fire retardants, painting inks, and other items, cyanides are also utilized as a raw material. They are utilized directly in a number of procedures, including electroplating and gold and silver extraction using hydrometallurgical methods. Currently, cyanide is created as a by-product in the aluminium smelting, chemical manufacturing, petroleum refining, and iron and steel industries. Cyanogenic substances are particularly harmful if unintentionally consumed by mammals or hydrolyzed by plants. This is mostly because of their capacity to decouple mitochondrial cytochrome C oxidase. HCN may easily form a stable and irreversible link with the Fe in cytochrome. These cause the electron transport chain to be disrupted, limiting the aerobic respiratory pathway, which accounts for 95% of the energy produced in cells in the form of ATP. Animal tissues that rely heavily on aerobic respiration for energy, such as the heart and central nervous system, are severely damaged. Therefore, if ATP synthesis is prevented, plants or animals would die fast because there won't be enough energy left over to perform their daily tasks.

Despite the risk of explosion when exposed to huge amounts of these explosives, such as TNT, exposure can have extremely harmful consequences on the body, including altered liver function, anemia, skin irritation, and cataracts. Similar to RDX, when inhaled or consumed in excessive amounts, it causes acute problems. TNT and RDX have long-term health impacts that can include heart and nervous system failure, which can be fatal. These hazardous materials may occasionally seep into the groundwater and cause toxicity distant from the affected locations. TNT and RDX have a variety of impacts on plants that are growing in polluted locations. These consequences include a delay in seed germination, growth and development, as well as the induction of leaf chlorosis and necrosis of plant tissues. The compounds can cause changes in both plants and animals since they are mutagenic. The severity of the negative effects of salt stress on crops or other naturally growing plants varies greatly depending on the species or cultivar, growth stage, and interaction with other environmental constraints. The most salt-tolerant Brassica species were discovered to be *B. napus*, whereas *B. campestris* and *B. nigra* were shown to be the most salt-susceptible.

Different strategies are used by plants to combat high soil salt concentrations. Osmoregulation, segregation of hazardous ions, ion excretion, scavenging of reactive oxygen species, the buildup of suitable solutes, etc. are a few of these. By delaying germination or maturity until favorable conditions, salt exclusion at the root level, or preferential root growth in non-saline areas, compartmentation of salts in vacuoles or specialized cells such as salt glands and salt hairs, or storage in older leaves, and selective discrimination of Na+ against K+ or Ca^{2+} are ways that plants can tolerate high ion concentrations. The antioxidant defence mechanism consists of antioxidant substances like tocopherols and carotenoids and enzymes such as superoxide dismutase (SOD), catalase (CAT), peroxidase (POD), etc. (Bhaduri and Fulekar, 2012). The capacity of plants to scavenge ROS varies. SOD, for instance, can catalyze the dismutation of superoxide into dioxygen and hydrogen peroxide in plants. H_2O_2 can be neutralized by catalases or peroxidase. Important factors in a plant's ability to tolerate salt are the accumulation of suitable solutes like polyols, sugars, glycine-betaine, proline, and other free amino acids. These solutes protect proteins from the salt-induced dissociation of their separate subunits while allowing cells to alter the osmotic potential to a level in the cytoplasm to maintain an adequate quantity of water content. In photosynthetic organisms, these organic solutes are essential for maintaining photosystem II integrity at high salt concentrations as well as the activity of photosynthesis-related enzymes like ribulose 1,5-bisphosphate carboxylase/oxygenase.

Even though some metals, like copper and zinc, serve as necessary components at low concentrations, they may turn harmful if they build up to higher amounts in the environment. Even in very low quantities, some metals (non-essential) can be hazardous to organisms. With the exception of some elements like selenium, iodine, and technetium, the concentration of essential elements in organisms is typically controlled homeostatically, meaning they are taken up from the environment in accordance with a plant's nutritional demand. The symptoms of toxicity or deficiency in organisms appear if this regulatory mechanism malfunctions due to either an inadequate supply (deficiency) or an excessive amount (toxicity) of metal. The levels of other significant metals and complex chemicals present in waters from various sources, the absorbing organism, and the physico-chemical properties of the soil environment, all play a role in the differential variability of uptake of various metals. Free ions are generally accessible forms of metals, and the amount of free ions present is typically a sign of potential toxicity. Valency greatly affects its bioavailability and mobility in soil and plants. Different mechanisms may have evolved in different species to survive high metal concentrations. In the majority of research, only one or a few metals are used to examine a plant tolerance capacity. However, most of the sites are polluted with multiple types of pollutants (organic and inorganic) with varied levels of toxicity when exposed to the elements. Other environmental and geophysical characteristics also significantly contribute to their accessibility and uptake.

8.8 METAL STRESS

The content of organic molecules and amino acids has been found to significantly rise in the roots of plants under metal stress. Some findings imply that reduced metal distribution from roots to shoots and thus decreased toxicity in aerial parts may be caused by the complexation of metals with these organic compounds and amino acids. However, there needs to be a reliable mechanism to lessen metals' toxicity once they reach the aerial regions. The most likely route of cellular sequestration is the compartmentalization of metal ions in the vacuole. Furthermore, the majority of metals cause the production of reactive oxygen species. Because of this, the majority of plants have developed an efficient scavenging system that includes both enzymatic and non-enzymatic antioxidants, including proline, ascorbic acid, tocopherols, glutathione, carotenoids, and phenolics. Enzymatic antioxidants include superoxide dismutase, peroxidase, catalase, glutathione reductase, and ascorbate reductase. These antioxidants scavenge reactive oxygen species and guard against oxidative damage to micro- and macromolecules as well as other cellular structures. Pesticides and herbicides have various effects on plants and animals. Some of these compounds have selective activity, whereas others have a broad spectrum of effects. Therefore, compared to selective pesticides, broad-range pesticides are more dangerous to the environment and species.

Numerous non-target organisms, including beneficial soil bacteria, crops, animals, and other aquatic species, are seriously threatened by pesticide contamination. To ensure the survival of these non-target species, hazardous substances must be avoided or used as little as. In soils with minerals and organic matter, many pesticides may interact with the soil and leave harmful residues that may not be removed from the soil even after extensive extraction. The toxicity of these bonded residues to plants and microorganisms is greatly influenced by their bioavailability. Even though it has been shown that the activities of soil microorganisms depend primarily on the release of bound residues from the soil, other factors, such as agronomic practises and the application of some other chemicals that may change the chemical nature of the soil, may also result in the release of soil-bound residues. This might lead to the compounds being recycled into the soil solution, where they could eventually be absorbed by plants and cause severe toxicity in plants. Numerous toxicity symptoms in plants have been linked to the overuse of pesticides and herbicides. However, depending on the type of chemical, active ingredient, and concentration in the developing environment, there is a wide range of toxicity symptoms. Inhibition of seed germination, growth retardation, loss of photosynthetic pigments, damages to the photosynthetic machinery, fruit drop, decreased yield, and a variety of other symptoms are the most typical toxicity symptoms in non-target plants.

8.9 BIOREMEDIATION OF EXPLOSIVES

Infection by pathogens, growth stage, herbivory, and environmental factors all have a significant impact on the quantities of cyanogenic glycosides. Plants typically compartmentalize cyanogenic glycosides in cell vacuoles, protecting cells from their toxicity. Thus, cyanogenic glycosides found in plant tissues are not harmful unless they are digested to generate free HCN by plant enzymes or rumen microbes. The cytoplasmic enzyme glucosidase is often responsible for this hydrolysis. Alanine and asparagine are two amino acids that can be converted into cyanogenic glycosides, which are compounds in which a simple sugar is joined to a cyanide molecule, which can then be used to detoxify cyanide. Typically, the quantity of nitro groups in an explosive is what determines how dangerous it is. Different plants may be able to absorb and break down dangerous explosives like nitroglycerine into less dangerous, simpler chemicals. Most recently, it has been demonstrated that plants that have undergone genetic engineering are better able to detoxify toxic substances. TNT and other hazardous explosives are transformed into various chemicals in these factories for further processing by plant enzymes. Many genes are up-regulated in response to the explosive presence, including transferases, which change the acceptor molecule bioactivity, solubility, and/or transport characteristics by adding a specific residue. It is possible to achieve phytoremediation of polluted soils using a variety of methods.

Heavy metals and salts are usually removed from contaminated soils using this technique. The long-term stability of the harmful contaminants is called phytostabilization. By creating an area around their roots where contaminants can be precipitated and fixed, some plants have the innate potential to immobilize toxins. Phytoremediation is nevertheless criticized because of some of its drawbacks. For instance, it can only reclaim soil that is at the surface or up to the depth. Without completely removing the contaminants from the soil, using plant-based remediation techniques makes it difficult to stop the leaching of pollutants into groundwater aquifers.

It has been suggested that an affordable method for treating saline soils is to use plants called halophytes that are very salt tolerant. Some halophytic species including those of Atriplex, Suaeda, Salsola, Chenopodium, and Portulaca have the capacity to absorb salt ions through roots and metabolize or store them through the process of phytoextraction in the leaves. By absorbing and storing salt, these halophytes can lower the salt content of the soil, at least at the rhizospheric level, and improve the soil's suitability for the growth of agricultural crops.

This strategy appears to be successful since many halophytic and salt-tolerant plant species can be used to reclaim saline soils because they naturally flourish on saline soils. When advanced biochemical and genetic modification methods and conventional soil reclamation are expensive, this strategy appears to be more affordable. The ability to tolerate salt, however, differs significantly between species as well as between populations of the same species.

The methods for removing heavy metals from polluted soils without affecting the fertility and structure of the soil are phytoextraction and phytoaccumulation. This method involves the accumulation of harmful metals into biomass that may be readily harvested from contaminated locations. Continuous or natural phytoextraction is the removal of metals based on the natural propensity to absorb metal pollutants without manifesting major toxicological effects. Contaminants are taken up by plants during phytovolatilization, converted into volatile compounds, and then released into the atmosphere through transpiration.

8.10 RHIZOFILTRATION

Lead, cadmium, copper, nickel, zinc, and chromium, which are predominantly kept in the roots, can be removed by rhizofiltration, which is the removal of metals by flowing through a mass of roots. Different crop plants, including sunflower, Indian mustard, tobacco, rye, spinach, and corn, as well as tree plants, including poplar, have been used. The ability of sunflower and poplar to remove metals from the contaminated environment is the strongest among these. Crops can also be used for various agricultural purposes, transformed into biofuel, or used in place of fossil fuel. However, the scope of this approach is quite constrained. First of all, the chosen plant species might perform poorly in severely contaminated sites but flourish in moderately contaminated areas. Second, this method will not be able to collect toxins from deep soil below the rooting depth. As a result, shallow-rooted plants won't be as effective as deep-rooted ones. Thirdly, it typically takes many years to bring the pollutant concentration down to acceptable levels. Many different pollutants, including metals, inorganics, and organics, are present at most sites. Rhizofiltration, which uses plants to remove pollutants, will not be sufficient in this situation and will need to be supported by other techniques. The health of animals and people may be threatened by plants that are grown in contaminated soil and water. In order to properly remediate soil and water using the rhizofiltration approach, special attention should be paid during harvest and only non-fodder crops should be selected (Subhashini and Swamy, 2017).

The majority of heavy metal pollution comes from an increasing number of anthropogenic sources. New methods, including phytoremediation, are being created as a result of the rise in heavy metal pollution of the environment. Plant species that are hyperaccumulators can be produced through genetic engineering techniques. When plants absorb too many heavy metals from soil solutions, a variety of cellular interactions occur that have toxic effects on the enzyme activity, protein structure, mineral nutrition, water balance, respiration and ATP content, photosynthesis, growth and morphogenesis, and production of reactive oxygen species (ROS). The relationship between a genotype and its environment causes a plant to manifest in the increased plant internal sequestration mechanism. Plants have metabolic machinery that helps them accumulate and tolerate heavy metals. Metal ion homeostasis and transportation are performed by metal transporters. Some organic and amino acids are ligands for heavy metals, and these organic and amino acids are crucial for detoxification and tolerance. Heavy metal stress causes plants to develop phytochelatins (PCs), which play a role in binding the metals to complexes and salts and securing the compounds inside the cell so that the metals cannot interfere with cell metabolism.

Heavy metal concentrations in plants frequently exceed those in soil by several folds, from where they enter the food chain. The ability of plants to absorb these metals and withstand extremely high concentrations is a feature that differs between species. Fast-growing plants with high biomass that can withstand or accumulate a variety of heavy metals in their harvestable portions are appropriate for phytoremediation. Heavy metal ions interact with the functional groups of proteins, nucleic acids, and polysaccharides when bound on the cell surface and within cells, and they take the place of other metal ions that are already bound to these functional groups. As different metabolic problems develop, it is frequently difficult to distinguish between those that are main and those that are secondary. Many heavy metals have a considerable binding affinity for ligands that contain sulfur. These interact with proteins as soon as they get inside the cell, altering the protein's original conformational shape. When they interact with enzymes, they hide the active site and interfere with its function. Since the environment is contaminated with heavy metals, technology is required to clean it using appropriate procedures. These techniques must be simple to use, affordable, and practical.

Among these technologies, phytoremediation has attracted the greatest attention because of its affordability, viability, and ease of use. Plants are living things that are subject to a variety of challenges, including air pollution, drought, heat, light, heavy metals, salt, cold, UV radiation, and nutrient shortage. A variety of cellular interactions may be the cause of the toxic symptoms seen in plants when high levels of heavy metals are present. In order to detect harmful heavy metals in the environment, plant development is frequently inhibited. Inhibition of growth by heavy metals arises from metabolic problems and direct impacts on growth, such as the reduction of cell wall flexibility caused by interactions with cell wall polysaccharides. It is critical to identify the lowest concentration at which root growth is inhibited in order to evaluate the ecological effects of heavy metals.

At moderate concentrations, the primary root length reduces more slowly than the number of lateral roots, and the root system develops a denser pattern. Due to the endodermal barrier and the unique cell structure in the central cylinder, the start of lateral roots is highly resistant to heavy metals. When heavy metals reduce the final size of elongated cells and consequently the spacing between lateral root initials, denser root systems result. Different plant species exhibit different levels of tolerance to heavy metals in the same enzyme. Understanding the multiple ways that heavy metal exposure affects various facets of cell metabolism is dependent on the loss in enzyme activity. These ions can occasionally even increase enzyme activity. Because these activities decline after brief exposure to heavy metal ions, the direct stimulation of catalase, peroxidase, and superoxide dismutase has not been proven conclusively. It appears that oxidative stress increased the amounts of free radicals and peroxides in the cytoplasm, which elevated the activities of the stress-related enzymes. The rates of cell division and elongation fall in the roots exposed to heavy metals due to a variety of factors. Direct DNA binding, metal-induced abnormalities, the enlargement of the mitotic cycle, suppression of microtubule formation, decreased cell wall plasticity, and a decrease in the glutathione pool are some of these processes. Both cell division and elongation are frequently inhibited by different substances, and in this instance, the sensitivity of the two processes to the inhibitory agent is not noticeably different. When compared to other metals, Cd and Pb have harmful effects on cell division and elongation, while other stressors cause different mitotic problems. Different plant tissues and species will react differently to heavy metals depending on the degree of disorder present and their ability to produce molecules that bind to heavy metals, removing them from the active metabolism.

In addition to those that result from processes involved in routine metabolism, such as photosynthesis and respiration, there are numerous additional potential sources of ROS in plants. The interaction of several ROS-producing and ROS-scavenging systems determines the equilibrium between the steady-state levels of distinct ROS. Several proteins act as hydrogen peroxide and superoxide scavengers. These comprise, among others, the peroxiredoxin family of proteins, glutathione reductase (GR), ascorbate peroxidase (APOX), catalase, superoxide dismutase (SOD), and thioredoxin. Numerous other non-protein scavengers, including glutathione and intracellular ascorbate, are added to these protein antioxidants. Because several heavy metals are engaged in a variety of ROS-generating pathways, their intoxication causes oxidative stress. Through the root system, primary heavy metal ions enter plants from the soil. The carboxyl group of mucilage uronic acid, which is present at the covering of the root system, binds with heavy metal ions at the root surface, however, the capacity of mucilage to bind heavy metals varies depending on the metal. The pH of the soil solution, the amount of organic matter, and the concentrations of other ions in the soil affect the absorption rate. The production of less soluble compounds causes many metal salts to become less soluble in soil solution at higher pH values, which reduces their biological availability in the soil. The absorption is increased when synthetic chelating agents like EDTA are added to heavy metal-polluted soil, and this property can be utilized to clean the soil.

Ions found in the soil solution have a significant impact on how well specific plant tissues absorb heavy metals. Although no specific mechanism is understood, it is likely that other ions in the soil solution interact and compete with one another, reducing the biological availability of metal ions and reducing their uptake. The relationship between a genotype and its environment determines how the increased plant internal sequestration mechanism manifests in the plant. Some plants are resistant to metal stress because they have a variety of possible processes that might be involved in the detoxification of heavy metals. Binding to the cell wall, decreased metal uptake or efflux pumping at the plasma membrane, chelation of metals in the cytosol by a variety of ligands such as phytochelatins, metallothioneins, and metal-binding proteins, repair of stress-damaged proteins, and compartmentation of metals in the vacuole by tonoplast-located transporters are some of these mechanisms. The multilayer cortex acts as the second barrier and protects plants from the toxic effects of heavy metals by binding the majority of these ions in the cell wall, which appears to lessen the toxic effects of metal ions on other tissues. Except for radish, the endodermis often has a lower concentration of heavy metals than the cortex, and occasionally the cortex is completely devoid of them. The endodermis and cortex have nearly identical heavy metal contents at high external concentrations. Heavy metals enter the stele more quickly where the lateral roots pierce the endodermis.

Heavy metals are not present in the parenchyma at levels below the fatal threshold. These have, however, been observed in some plants in the apoplast of the basal root region and occasionally even in the vacuoles of root stellar cells. They pass through the cortex and endodermis at concentrations that are on the verge of being deadly and are primarily present in the vascular cylinder cells. Heavy metal penetration in the root meristem has not been sufficiently researched. Due to the unusual features of the cell wall and plasmalemma in these cells, it was nearly completely missing from the region of the dormant core. The distribution of heavy metals in the stem and leaf tissues of plants grown in metal solutions is a subject of very little research. The majority of metals are concentrated in the cortex and rhizodermis and do not breach the endodermal barrier at sublethal levels. The endodermal barrier is breached at deadly doses, allowing a flood of heavy metals to penetrate the inner tissues. The tissues on the root surface act as barriers to restrict uptake into the root. Understanding the mechanism of plant resistance is made possible by research into the endodermal barrier, which limits transport. Depending on the transport mechanism, they are distributed unevenly in plant tissues.

8.11 METALLOTHIONEINS

The majority of eukaryotes use complex formation for metal detoxification. Animals, higher plants, and some prokaryotes have cysteine-rich proteins known as metallothioneins (MTs), which have a low molecular weight. Hormones, cytotoxic substances, and metals like Cd, Zn, Hg, Cu, Au, Ag, Co, Ni, and Bi can increase the production of MTs, which is controlled at the transcriptional level. They perform the role of molecular chaperones in the normal folding and assembly of proteins. HSP is expressed in the roots of *Armeria maritima* plants that are grown in soils high in copper. A short heat stress administered before heavy metal stress was also reported to create a tolerance effect by preventing membrane damage. Plant nutrition and tolerance to harmful heavy metals depend on maintaining metal cation equilibrium (Rono et al., 2021). As a result, the study of heavy metal transport in plants is a very interesting and growing topic. Current studies have shown that plants have several classes of metal transporters that may be involved in metal uptake and homeostasis in general and, therefore, could be crucial for tolerance, despite the lack of direct evidence for a role for plasma membrane efflux transporters in heavy metal tolerance in plants. These include proteins from the cation diffusion facilitator (CDF) family, the zinc-iron permease (ZIP) family, the natural resistance-associated macrophage (Nramp) family, and heavy metal ATPases. Naturally, there are still a lot of plant metal transporters that need to be found at the molecular level.

Zn, Cd, and Co are transported via a family of heavy metal transporters known as the CDF proteins. While some members of the CD family are found in plasma membranes and others are found in intracellular membranes, some are assumed to function in heavy metal uptake while others catalyze efflux. Zinc sequestration may be aided by the Arabidopsis thaliana ZAT1 protein zinc transporter. Transgenic plants that overexpress ZAT1 were found to have improved zinc resistance, and when exposed to high zinc concentrations, roots contained more zinc. Northern blotting study revealed that ZAT1 was constitutively expressed throughout the plant body and was not stimulated by exposure to increasing quantities of zinc, indicating that this transporter is not restricted to root tissue. Theoretically, it matters a lot how plants differ from one another in their capacity to collect metals because this is the basis upon which natural selection acts to shape the evolution of hyperaccumulation. There is little information on the processes by which airborne contaminants are absorbed and accumulated, their permanence in moss and lichens, and the potential synergistic and/or antagonistic effects of climatic and environmental conditions. As some metals have synergistic toxicity and risk may exist even under low-dose exposure conditions, this can result in a serious underestimation of potential health impacts. Along with removing the chance of contamination through root uptake, bags are more effective and less expensive than volume samplers. The method's main drawback is its unknowable collection efficiency for various pollutants. The reported metal concentrations may thus represent relative rates of deposition rather than the total load of pollutants in the atmosphere.

Another crucial element of biomonitoring via bags is the exposure time. Climate and environmental factors may affect biomonitoring performance and biomonitors may approach a saturation point for the uptake of an element. The quantities of trace elements in the thallus are comparable to those in wet and dry air depositions and are simple to quantify when compared to instrumental monitoring. Despite the fact that many metal elements are necessary for plant growth at low concentrations, their excessive levels in soil that are over threshold values can be harmful. The nature of an element and the type of plant has an impact on this negative outcome. The reactions of plants to non-essential metals and metalloids, such as Hg, Cd, Cr, Pb, As, and Sb, range widely from tolerance to toxicity. These elements are harmful in their elemental and chemically mixed forms. For instance, the replacement of necessary metals in pigments or enzymes by high amounts of heavy metals may result in oxidative stress and the disruption of their function. The growth of subterranean and aboveground plant parts as well as the function of the photosynthetic apparatus are hampered by heavy metal toxicity, which is frequently associated with the progression of senescence. Plants have developed unique methods by which hazardous materials are removed, kept at the root level, or changed into physiologically tolerant forms in order to avoid toxicity.

The definition of heavy metal is an element with a density of more than 5 g cm^3. While chromium (Cr) and antimony (Sb) are known to be essential for animals, other heavy metals, including cobalt (Co), copper (Cu), iron (Fe), manganese (Mn), molybdenum (Mo), nickel (Ni), and zinc (Zn), are regarded to be needed for plants. Due to their high reactivity, these metal elements can directly affect the processes of growth, senescence, and energy production. Beyond acceptable limits, their concentration in soil is hazardous to plants, either by inducing oxidative stress through the production of free radicals or by impairing enzyme activity by displacing vital metals and minerals. Although adjustments in cell metabolism help plants adapt, the primary sign of metal poisoning is a decline in plant growth. On the other hand, the adaptation mechanisms in older plants that have been exposed to heavy metals are less adaptable and effective. As a result, there are considerably more obvious harmful effects of heavy metals on the physiology and metabolism of plants.

Chromium and cadmium are two heavy metals that are particularly problematic since they have the potential to be unsafe to both plants and animals, even at low doses. The effects of chromium toxicity on plants range from mutagenesis to enzymatic activity suppression. Stunting, yield decline, and leaf chlorosis are some of the indicators that are. Due to its high toxicity and significant solubility in water, cadmium (Cd) is a particularly hazardous contaminant. Cd interferes with the ability to absorb metal nutrients like Fe, Zn, Cu, and Mn, as well as causing plants to break down their chlorophyll and experience lipid peroxidation, which increases the production of reactive oxygen species (ROS). Because it utilizes the same transmembrane carriers as K, Ca, Mg, and Fe, cadmium likewise prevents the uptake of these elements. Its buildup in plants may also provide a major health danger to people through the food chain, but it also poses an additional risk to children by ingesting Cd-contaminated soil directly. Heavy metals that enter our environment are transported by water and air before being dumped in soil and sediments where they may become immobile. The bonding process, however, could take a very long time. It has been observed that the bioavailable percentage of metal elements in the soil is high at the start of the binding process but steadily declines over time. Chemical characteristics of soil, such as soil pH, loading rate, cation exchange capacity, redox potential, soil texture, clay content, and organic matter, have a significant impact on metal solubility and bioavailability to plants. The metals will typically be more tightly linked to soil with a longer residence period and less bioavailable to plants the higher the clay and/or organic matter and soil pH. In addition, soil temperature has a significant role in explaining changes in metal uptake by crops.

References

1. Agnello, A. C., Bagard, M., van Hullebusch, E. D., Esposito, G., & Huguenot, D. (2016). Comparative bioremediation of heavy metals and petroleum hydrocarbons co-contaminated soil by natural attenuation, phytoremediation, bioaugmentation and bioaugmentation-assisted phytoremediation. Science of The Total Environment, 563–564, 693–703. https://doi.org/10.1016/j.scitotenv.2015.10.061
2. Agostini, E., Coniglio, M. S., Milrad, S. R., Tigier, H. A., & Giulietti, A. M. (2003). Phytoremediation of 2,4-dichlorophenol by Brassica napus hairy root cultures. Biotechnology and Applied Biochemistry, 37(2), 139–144. https://doi.org/10.1042/BA20020079
3. Ahemad, M. (2015). Phosphate-solubilizing bacteria-assisted phytoremediation of metalliferous soils: A review. 3 Biotech, 5(2), 111–121. https://doi.org/10.1007/s13205-014-0206-0
4. AL Sbani, N. H., Abdullah, S. R. S., Idris, M., Hasan, H. A., Halmi, M. I. E., Jehawi, O. H., & Ismail, N. 'Izzati. (2021). PAH-degrading rhizobacteria of Lepironia articulata for phytoremediation enhancement. Journal of Water Process Engineering, 39, 101688. https://doi.org/10.1016/j.jwpe.2020.101688
5. Appenroth, K.-J., Krech, K., Keresztes, Á., Fischer, W., & Koloczek, H. (2010). Effects of nickel on the chloroplasts of the duckweeds Spirodela polyrhiza and Lemna minor and their possible use in biomonitoring and phytoremediation. Chemosphere, 78(3), 216–223. https://doi.org/10.1016/j.chemosphere.2009.11.007

6. Ashima Bhan, & N. N. Sarkar. (2005). Mercury in the Environment: Effect on Health and Reproduction. Reviews on Environmental Health, 20(1), 39–56. https://doi.org/10.1515/REVEH.2005.20.1.39
7. Aslam, S., & Gul, N. (2020). Impact of Genetically Modified Crops on Environment. In R. M. Singh, P. Shukla, & P. Singh (Eds.), Environmental Processes and Management: Tools and Practices (pp. 237–248). Cham: Springer International Publishing. https://doi.org/10.1007/978-3-030-38152-3_13
8. Athapattu, B. C. L., Thalgaspitiya, T. W. L. R., Yasaratne, U. L. S., & Vithanage, M. (2016). Phytoremediation Using Constructed Wetlands to Treat Reverse Osmosis Rejected Concentrates. Proceedings of International Forestry and Environment Symposium, 21. https://doi.org/10.31357/fesympo.v21i0.3227
9. Awa, S. H., & Hadibarata, T. (2020). Removal of Heavy Metals in Contaminated Soil by Phytoremediation Mechanism: A Review. Water, Air, & Soil Pollution, 231(2), 47. https://doi.org/10.1007/s11270-020-4426-0
10. B. Arnao, M., & Hernández-Ruiz, J. (2019). Role of Melatonin to Enhance Phytoremediation Capacity. Applied Sciences, 9(24), 5293. https://doi.org/10.3390/app9245293
11. Bahraminia, M., Zarei, M., Ronaghi, A., & Ghasemi-Fasaei, R. (2016). Effectiveness of arbuscular mycorrhizal fungi in phytoremediation of lead- contaminated soil by vetiver grass. International Journal of Phytoremediation, 18(7), 730–737. https://doi.org/10.1080/15226514.2015.1131242
12. Banerjee, A., Cardozo, S., Khuntia, S. P., & Upadhyay, A. K. (2022). Chapter 19—Phytoremediation of pharmaceutical wastes. In R. A. Bhat, F. M. P. Tonelli, G. H. Dar, & K. Hakeem (Eds.), Phytoremediation (pp. 387–414). Academic Press. https://doi.org/10.1016/B978-0-323-89874-4.00003-0
13. Banerjee, R., Goswami, P., Lavania, S., Mukherjee, A., & Lavania, U. C. (2019). Vetiver grass is a potential candidate for phytoremediation of iron ore mine spoil dumps. Ecological Engineering, 132, 120–136. https://doi.org/10.1016/j.ecoleng.2018.10.012
14. Barocsi, A., Csintalan, Z., Kocsanyi, L., Dushenkov, S., Kuperberg, J. M., Kucharski, R., & Richter, P. I. (2003). Optimizing Phytoremediation of Heavy Metal-Contaminated Soil by Exploiting Plants' Stress Adaptation. International Journal of Phytoremediation, 5(1), 13–23.

https://doi.org/10.1080/16226510390856448

15. Batty, L. C., & Anslow, M. (2008). Effect of a Polycyclic Aromatic Hydrocarbon on the Phytoremediation of Zinc by Two Plant Species (Brassica juncea and Festuca arundinacea). International Journal of Phytoremediation, 10(3), 236–251. https://doi.org/10.1080/15226510801997549
16. Bhaduri, A. M., & Fulekar, M. H. (2012). Antioxidant enzyme responses of plants to heavy metal stress. Reviews in Environmental Science and Bio/Technology, 11(1), 55–69. https://doi.org/10.1007/s11157-011-9251-x
17. Biswal, T. (2022). Genetic Systems in Soil Bacteria for Degradation of Polychlorinated Biphenyls. In Bioremediation and Phytoremediation Technologies in Sustainable Soil Management. Apple Academic Press.
18. Bizily, S. P., Rugh, C. L., Summers, A. O., & Meagher, R. B. (1999). Phytoremediation of methylmercury pollution: MerB expression in Arabidopsis thaliana confers resistance to organomercurials. Proceedings of the National Academy of Sciences, 96(12), 6808–6813. https://doi.org/10.1073/pnas.96.12.6808
19. Bokhari, S. H., Ahmad, I., Mahmood-Ul-Hassan, M., & Mohammad, A. (2016). Phytoremediation potential of Lemna minor L. for heavy metals. International Journal of Phytoremediation, 18(1), 25–32. https://doi.org/10.1080/15226514.2015.1058331
20. Bonanno, G. (2013). Comparative performance of trace element bioaccumulation and biomonitoring in the plant species Typha domingensis, Phragmites australis and Arundo donax. Ecotoxicology and Environmental Safety, 97, 124–130. https://doi.org/10.1016/j.ecoenv.2013.07.017
21. Brandt, R., Merkl, N., Schultze-Kraft, R., Infante, C., & Broll, G. (2006). Potential of vetiver (Vetiveria zizanioides (L.) Nash) for phytoremediation of petroleum hydrocarbon-contaminated soils in Venezuela. International Journal of Phytoremediation, 8(4), 273–284. https://doi.org/10.1080/15226510600992808
22. Burken, J. G. (2003). Uptake and Metabolism of Organic Compounds: Green-Liver Model. In Phytoremediation (pp. 59–84). John Wiley & Sons, Ltd. https://doi.org/10.1002/047127304X.ch2
23. Burken, Joel G., & Schnoor, J. L. (1996). Phytoremediation: Plant Uptake of Atrazine and Role of Root Exudates. Journal of Environmental

Engineering, 122(11), 958–963. https://doi.org/10.1061/(ASCE)0733-9372(1996)122:11(958)

24. Carvalho, P. N. (2021). Constructed Wetlands and Phytoremediation as a Tool for Pharmaceutical Removal. In S. Pérez Solsona, N. Montemurro, S. Chiron, & D. Barceló (Eds.), Interaction and Fate of Pharmaceuticals in Soil-Crop Systems: The Impact of Reclaimed Wastewater (pp. 377–413). Cham: Springer International Publishing. https://doi.org/10.1007/698_2020_624
25. Chandra, R., & Kumar, V. (2017a). Phytoextraction of heavy metals by potential native plants and their microscopic observation of root growing on stabilised distillery sludge as a prospective tool for in situ phytoremediation of industrial waste. Environmental Science and Pollution Research, 24(3), 2605–2619. https://doi.org/10.1007/s11356-016-8022-1
26. Chandra, R., & Kumar, V. (2017b). Phytoremediation: A Green Sustainable Technology for Industrial Waste Management. In Phytoremediation of Environmental Pollutants. CRC Press.
27. Chandra, R., Kumar, V., & Singh, K. (2017). Hyperaccumulator versus Nonhyperaccumulator Plants for Environmental Waste Management. In Phytoremediation of Environmental Pollutants. CRC Press.
28. Chaney, R. L., Broadhurst, C. L., & Centofanti, T. (2010). Phytoremediation of Soil Trace Elements. In Trace Elements in Soils (pp. 311–352). John Wiley & Sons, Ltd. https://doi.org/10.1002/9781444319477.ch14
29. Chen, H., Xiong, J., Fang, L., Han, F., Zhao, X., Fan, Q., & Tan, W. (2022). Sequestration of heavy metals in soil aggregates induced by glomalin-related soil protein: A five-year phytoremediation field study. Journal of Hazardous Materials, 437, 129445. https://doi.org/10.1016/j.jhazmat.2022.129445
30. Chen, W.-C., Juang, R.-S., & Wei, Y.-H. (2015). Applications of a lipopeptide biosurfactant, surfactin, produced by microorganisms. Biochemical Engineering Journal, 103, 158–169. https://doi.org/10.1016/j.bej.2015.07.009
31. Chen, Y., Shen, Z., & Li, X. (2004). The use of vetiver grass (Vetiveria zizanioides) in the phytoremediation of soils contaminated with heavy metals. Applied Geochemistry, 19(10), 1553–1565. https://doi.org/10.1016/j.apgeochem.2004.02.003

32. Collins, C., Laturnus, F., & Nepovim, A. (2002). Remediation of BTEX and trichloroethene. Environmental Science and Pollution Research, 9(1), 86–94. https://doi.org/10.1007/BF02987319

33. Conger, R. M., & Portier, R. J. (1997). Phytoremediation experimentation with the herbicide bentazon. Remediation Journal, 7(2), 19–37. https://doi.org/10.1002/rem.3440070203

34. DalCorso, G., Farinati, S., Maistri, S., & Furini, A. (2008). How Plants Cope with Cadmium: Staking All on Metabolism and Gene Expression. Journal of Integrative Plant Biology, 50(10), 1268–1280. https://doi.org/10.1111/j.1744-7909.2008.00737.x

35. Das, S., Goswami, S., & Talukdar, A. D. (2014). A Study on Cadmium Phytoremediation Potential of Water Lettuce, Pistia stratiotes L. Bulletin of Environmental Contamination and Toxicology, 92(2), 169–174. https://doi.org/10.1007/s00128-013-1152-y

36. De Conti, L., Ceretta, C. A., Melo, G. W. B., Tiecher, T. L., Silva, L. O. S., Garlet, L. P., … Brunetto, G. (2019). Intercropping of young grapevines with native grasses for phytoremediation of Cu-contaminated soils. Chemosphere, 216, 147–156. https://doi.org/10.1016/j.chemosphere.2018.10.134

37. Del Buono, D., Terzano, R., Panfili, I., & Bartucca, M. L. (2020). Phytoremediation and detoxification of xenobiotics in plants: Herbicide-safeners as a tool to improve plant efficiency in the remediation of polluted environments. A mini-review. International Journal of Phytoremediation, 22(8), 789–803. https://doi.org/10.1080/15226514.2019.1710817

38. Deng, Z., & Cao, L. (2017). Fungal endophytes and their interactions with plants in phytoremediation: A review. Chemosphere, 168, 1100–1106. https://doi.org/10.1016/j.chemosphere.2016.10.097

39. Deng, Z., Wang, W., Tan, H., & Cao, L. (2012). Characterization of Heavy Metal-Resistant Endophytic Yeast Cryptococcus sp. CBSB78 from Rapes (Brassica chinensis) and Its Potential in Promoting the Growth of Brassica spp. In Metal-Contaminated Soils. Water, Air, & Soil Pollution, 223(8), 5321–5329. https://doi.org/10.1007/s11270-012-1282-6

40. Dhankher, O. P., Pilon-Smits, E. A. H., Meagher, R. B., & Doty, S. (2012). 20—Biotechnological approaches for phytoremediation. In A. Altman & P. M. Hasegawa (Eds.), Plant Biotechnology and Agriculture

(pp. 309–328). San Diego: Academic Press. https://doi.org/10.1016/B978-0-12-381466-1.00020-1

41. Di Gregorio, S., Giorgetti, L., Ruffini Castiglione, M., Mariotti, L., & Lorenzi, R. (2015). Phytoremediation for improving the quality of effluents from a conventional tannery wastewater treatment plant. International Journal of Environmental Science and Technology, 12(4), 1387–1400. https://doi.org/10.1007/s13762-014-0522-2
42. Djedidi, S., Kojima, K., Yamaya, H., Ohkama-Ohtsu, N., Bellingrath-Kimura, S. D., Watanabe, I., & Yokoyama, T. (2014). Stable cesium uptake and accumulation capacities of five plant species as influenced by bacterial inoculation and cesium distribution in the soil. Journal of Plant Research, 127(5), 585–597. https://doi.org/10.1007/s10265-014-0647-x
43. Dudai, N., Tsion, I., Shamir, S. Z., Nitzan, N., Chaimovitsh, D., Shachter, A., & Haim, A. (2018). Agronomic and economic evaluation of Vetiver grass (Vetiveria zizanioides L.) as means for phytoremediation of diesel polluted soils in Israel. Journal of Environmental Management, 211, 247–255. https://doi.org/10.1016/j.jenvman.2018.01.013
44. Dunn, C. E., Brooks, R. R., Edmondson, J., Leblanc, M., & Reeves, R. D. (1996). Biogeochemical studies of metal-tolerant plants from southern Morocco. Journal of Geochemical Exploration, 56(1), 13–22. https://doi.org/10.1016/0375-6742(96)00007-6
45. Dzantor, E. K. (2007). Phytoremediation: The state of rhizosphere 'engineering' for accelerated rhizodegradation of xenobiotic contaminants. Journal of Chemical Technology & Biotechnology, 82(3), 228–232. https://doi.org/10.1002/jctb.1662
46. Ekperusi, A. O., Sikoki, F. D., & Nwachukwu, E. O. (2019). Application of common duckweed (Lemna minor) in phytoremediation of chemicals in the environment: State and future perspective. Chemosphere, 223, 285–309. https://doi.org/10.1016/j.chemosphere.2019.02.025
47. Entry, J. A., Vance, N. C., Hamilton, M. A., Zabowski, D., Watrud, L. S., & Adriano, D. C. (1996). Phytoremediation of soil contaminated with low concentrations of radionuclides. Water, Air, and Soil Pollution, 88(1), 167–176. https://doi.org/10.1007/BF00157420
48. Freitas, M. N. V. P. and H. (2005). Metal-Tolerant Plants: Biodiversity Prospecting for Phytoremediation Technology. In Trace Elements in the Environment. CRC Press.

49. Gatheru Waigi, M., Sun, K., & Gao, Y. (2017). Sphingomonads in Microbe-Assisted Phytoremediation: Tackling Soil Pollution. Trends in Biotechnology, 35(9), 883–899. https://doi.org/10.1016/j.tibtech.2017.06.014
50. Gavrilescu, M. (2022). Enhancing phytoremediation of soils polluted with heavy metals. Current Opinion in Biotechnology, 74, 21–31. https://doi.org/10.1016/j.copbio.2021.10.024
51. Gerhardt, K. E., Gerwing, P. D., & Greenberg, B. M. (2017). Opinion: Taking phytoremediation from proven technology to accepted practice. Plant Science, 256, 170–185. https://doi.org/10.1016/j.plantsci.2016.11.016
52. Glick, B. R. (2020). Phytoremediation. In B. R. Glick (Ed.), Beneficial Plant-Bacterial Interactions (pp. 319–359). Cham: Springer International Publishing. https://doi.org/10.1007/978-3-030-44368-9_10
53. Grobelak, A., & Hiller, J. (2017). Bacterial siderophores promote plant growth: Screening of catechol and hydroxamate siderophores. International Journal of Phytoremediation, 19(9), 825–833. https://doi.org/10.1080/15226514.2017.1290581
54. Guidi Nissim, W., & Labrecque, M. (2021). Reclamation of urban brownfields through phytoremediation: Implications for building sustainable and resilient towns. Urban Forestry & Urban Greening, 65, 127364. https://doi.org/10.1016/j.ufug.2021.127364
55. Gupta, D. K., Huang, H. G., & Corpas, F. J. (2013). Lead tolerance in plants: Strategies for phytoremediation. Environmental Science and Pollution Research, 20(4), 2150–2161. https://doi.org/10.1007/s11356-013-1485-4
56. Hannink, N. K., Rosser, S. J., & Bruce, N. C. (2002). Phytoremediation of Explosives. Critical Reviews in Plant Sciences, 21(5), 511–538. https://doi.org/10.1080/0735-260291044340
57. Haritash, A. K., & Kaushik, C. P. (2009). Biodegradation aspects of Polycyclic Aromatic Hydrocarbons (PAHs): A review. Journal of Hazardous Materials, 169(1), 1–15. https://doi.org/10.1016/j.jhazmat.2009.03.137
58. Harvey, P. J., Campanella, B. F., Castro, P. M. L., Harms, H., Lichtfouse, E., Schäffner, A. R., … Werck-Reichhart, D. (2002). Phytoremediation of polyaromatic hydrocarbons, anilines and phenols. Environmental Science and Pollution Research, 9(1), 29–47. https://doi.org/10.1007/BF02987315

59. Herath, I., & Vithanage, M. (2015). Phytoremediation in Constructed Wetlands. In A. A. Ansari, S. S. Gill, R. Gill, G. R. Lanza, & L. Newman (Eds.), Phytoremediation: Management of Environmental Contaminants, Volume 2 (pp. 243–263). Cham: Springer International Publishing. https://doi.org/10.1007/978-3-319-10969-5_21

60. Hoagland, R. E., Zablotowicz, R. M., & Locke, M. A. (1997). An Integrated Phytoremediation Strategy for Chloroacetamide Herbicides in Soil. In ACS Symposium Series: Vol.664. Phytoremediation of Soil and Water Contaminants (Vol. 664, pp. 92–105). American Chemical Society. https://doi.org/10.1021/bk-1997-0664.ch007

61. Hopmans, J. W., Qureshi, A. S., Kisekka, I., Munns, R., Grattan, S. R., Rengasamy, P., … Taleisnik, E. (2021). Chapter One—Critical knowledge gaps and research priorities in global soil salinity. In D. L. Sparks (Ed.), Advances in Agronomy (Vol. 169, pp. 1–191). Academic Press. https://doi.org/10.1016/bs.agron.2021.03.001

62. Hussain, S., Siddique, T., Arshad, M., & Saleem, M. (2009). Bioremediation and Phytoremediation of Pesticides: Recent Advances. Critical Reviews in Environmental Science and Technology, 39(10), 843–907. https://doi.org/10.1080/10643380801910090

63. Hwang, J.-U., Song, W.-Y., Hong, D., Ko, D., Yamaoka, Y., Jang, S., … Lee, Y. (2016). Plant ABC Transporters Enable Many Unique Aspects of a Terrestrial Plant's Lifestyle. Molecular Plant, 9(3), 338–355. https://doi.org/10.1016/j.molp.2016.02.003

64. I. Alkorta, J. M. Becerril, & C. Garbisu. (2010). Phytostabilization of Metal Contaminated Soils. Reviews on Environmental Health, 25(2), 135–146. https://doi.org/10.1515/REVEH.2010.25.2.135

65. Ibrahim, N., & El Afandi, G. (2020). Phytoremediation uptake model of heavy metals (Pb, Cd and Zn) in soil using Nerium oleander. Heliyon, 6(7), e04445. https://doi.org/10.1016/j.heliyon.2020.e04445

66. Jesus, J. M., Danko, A. S., Fiúza, A., & Borges, M.-T. (2015). Phytoremediation of salt-affected soils: A review of processes, applicability, and the impact of climate change. Environmental Science and Pollution Research, 22(9), 6511–6525. https://doi.org/10.1007/s11356-015-4205-4

67. Jisha, C. K., Bauddh, K., & Shukla, S. K. (2017). Phytoremediation and Bioenergy Production Efficiency of Medicinal and Aromatic Plants. In K. Bauddh, B. Singh, & J. Korstad (Eds.), Phytoremediation Potential

of Bioenergy Plants (pp. 287–304). Singapore: Springer. https://doi.org/10.1007/978-981-10-3084-0_11

68. Khan, H. N., & Faisal, M. (2018). Phytoremediation of Industrial Wastewater by Hydrophytes. In Abid A. Ansari, S. S. Gill, R. Gill, G. R. Lanza, & L. Newman (Eds.), Phytoremediation: Management of Environmental Contaminants, Volume 6 (pp. 179–200). Cham: Springer International Publishing. https://doi.org/10.1007/978-3-319-99651-6_8
69. Kidd, P., Barceló, J., Bernal, M. P., Navari-Izzo, F., Poschenrieder, C., Shilev, S., … Monterroso, C. (2009). Trace element behaviour at the root–soil interface: Implications in phytoremediation. Environmental and Experimental Botany, 67(1), 243–259. https://doi.org/10.1016/j.envexpbot.2009.06.013
70. Kösesakal, T., Ünal, M., Kulen, O., Memon, A., & Yüksel, B. (2016). Phytoremediation of petroleum hydrocarbons by using a freshwater fern species Azolla filiculoides Lam. International Journal of Phytoremediation, 18(5), 467–476. https://doi.org/10.1080/15226514.2015.1115958
71. Kotrba, P., Najmanova, J., Macek, T., Ruml, T., & Mackova, M. (2009). Genetically modified plants in phytoremediation of heavy metal and metalloid soil and sediment pollution. Biotechnology Advances, 27(6), 799–810. https://doi.org/10.1016/j.biotechadv.2009.06.003
72. Kotyza, J., Soudek, P., Kafka, Z., & Vaněk, T. (2010). Phytoremediation of Pharmaceuticals—Preliminary Study. International Journal of Phytoremediation, 12(3), 306–316. https://doi.org/10.1080/15226510903563900
73. Kurade, M. B., Ha, Y.-H., Xiong, J.-Q., Govindwar, S. P., Jang, M., & Jeon, B.-H. (2021). Phytoremediation as a green biotechnology tool for emerging environmental pollution: A step forward towards sustainable rehabilitation of the environment. Chemical Engineering Journal, 415, 129040. https://doi.org/10.1016/j.cej.2021.129040
74. Lenoir, I., Lounes-Hadj Sahraoui, A., & Fontaine, J. (2016). Arbuscular mycorrhizal fungal-assisted phytoremediation of soil contaminated with persistent organic pollutants: A review. European Journal of Soil Science, 67(5), 624–640. https://doi.org/10.1111/ejss.12375
75. Li, J., Meng, B., Chai, H., Yang, X., Song, W., Li, S., … Sun, W. (2019). Arbuscular Mycorrhizal Fungi Alleviate Drought Stress in C3 (Leymus chinensis) and C4 (Hemarthria altissima) Grasses via

Altering Antioxidant Enzyme Activities and Photosynthesis. Frontiers in Plant Science, 10. Retrieved from https://www.frontiersin.org/articles/10.3389/fpls.2019.00499

76. Licht, L. A., & Isebrands, J. G. (2005). Linking phytoremediated pollutant removal to biomass economic opportunities. Biomass and Bioenergy, 28(2), 203–218. https://doi.org/10.1016/j.biombioe.2004.08.015

77. Lombi, E., Zhao, F. J., Dunham, S. J., & McGrath, S. P. (2001). Phytoremediation of Heavy Metal–Contaminated Soils: Natural Hyperaccumulation versus Chemically Enhanced Phytoextraction. Journal of Environmental Quality, 30(6), 1919–1926. https://doi.org/10.2134/jeq2001.1919

78. Ma, Y., Rajkumar, M., Zhang, C., & Freitas, H. (2016a). Beneficial role of bacterial endophytes in heavy metal phytoremediation. Journal of Environmental Management, 174, 14–25. https://doi.org/10.1016/j.jenvman.2016.02.047

79. Ma, Y., Rajkumar, M., Zhang, C., & Freitas, H. (2016b). Inoculation of Brassica oxyrrhina with plant growth promoting bacteria for the improvement of heavy metal phytoremediation under drought conditions. Journal of Hazardous Materials, 320, 36–44. https://doi.org/10.1016/j.jhazmat.2016.08.009

80. Macek, T., Macková, M., & Káš, J. (2000). Exploitation of plants for the removal of organics in environmental remediation. Biotechnology Advances, 18(1), 23–34. https://doi.org/10.1016/S0734-9750(99)00034-8

81. Mackova, M., Barriault, D., Francova, K., Sylvestre, M., Möder, M., Vrchotova, B., … Macek, T. (2006). Phytoremediation of Polychlorinated Biphenyls. In M. Mackova, D. Dowling, & T. Macek (Eds.), Phytoremediation Rhizoremediation (pp. 143–167). Dordrecht: Springer Netherlands. https://doi.org/10.1007/978-1-4020-4999-4_11

82. Mahar, A., Wang, P., Ali, A., Awasthi, M. K., Lahori, A. H., Wang, Q., … Zhang, Z. (2016). Challenges and opportunities in the phytoremediation of heavy metals contaminated soils: A review. Ecotoxicology and Environmental Safety, 126, 111–121. https://doi.org/10.1016/j.ecoenv.2015.12.023

83. Malik, B., Pirzadah, T. B., & Hakeem, K. R. (2022). Chapter 20—Phytoremediation of persistent organic pollutants (POPs). In R. A. Bhat, F. M. P. Tonelli, G. H. Dar, & K. Hakeem (Eds.), Phytoremediation

(pp. 415–436). Academic Press. https://doi.org/10.1016/B978-0-323-89874-4.00010-8

84. Manoj, S. R., Karthik, C., Kadirvelu, K., Arulselvi, P. I., Shanmugasundaram, T., Bruno, B., & Rajkumar, M. (2020). Understanding the molecular mechanisms for the enhanced phytoremediation of heavy metals through plant growth promoting rhizobacteria: A review. Journal of Environmental Management, 254, 109779. https://doi.org/10.1016/j.jenvman.2019.109779
85. Manousaki, E., & Kalogerakis, N. (2011). Halophytes—An Emerging Trend in Phytoremediation. International Journal of Phytoremediation, 13(10), 959–969. https://doi.org/10.1080/15226514.2010.532241
86. Mant, C., Costa, S., Williams, J., & Tambourgi, E. (2006). Phytoremediation of chromium by model constructed wetland. Bioresource Technology, 97(15), 1767–1772. https://doi.org/10.1016/j.biortech.2005.09.010
87. Mata, T. M., Rodrigues, S., Caetano, N. S., & Martins, A. A. (2022). Life cycle assessment of bioethanol from corn stover from soil phytoremediation. Energy Reports, 8, 468–474. https://doi.org/10.1016/j.egyr.2022.01.059
88. Mawang, C.-I., Azman, A.-S., Fuad, A.-S. M., & Ahamad, M. (2021). Actinobacteria: An eco-friendly and promising technology for the bioaugmentation of contaminants. Biotechnology Reports, 32, e00679. https://doi.org/10.1016/j.btre.2021.e00679
89. McGrath, S. P., Zhao, F. J., & Lombi, E. (2001). Plant and rhizosphere processes involved in phytoremediation of metal-contaminated soils. Plant and Soil, 232(1), 207–214. https://doi.org/10.1023/A:1010358708525
90. McIntyre, T. (2003). Phytoremediation of Heavy Metals from Soils. In D. T. Tsao (Ed.), Phytoremediation (pp. 97–123). Berlin, Heidelberg: Springer. https://doi.org/10.1007/3-540-45991-X_4
91. McKnight, A. M., Gannon, T. W., & Yelverton, F. (2022). Phytoremediation of azoxystrobin and imidacloprid by wetland plant species Juncus effusus, Pontederia cordata and Sagittaria latifolia. International Journal of Phytoremediation, 24(2), 196–204. https://doi.org/10.1080/15226514.2021.1932726
92. Meagher, R. B. (2000). Phytoremediation of toxic elemental and organic pollutants. Current Opinion in Plant Biology, 3(2), 153–162. https://doi.org/10.1016/S1369-5266(99)00054-0

93. Mirzaee, M. M., ZakeriNia, M., & Farasati, M. (2021). The effects of phytoremediation of treated urban wastewater on the discharge of surface and subsurface drippers (Case study: Gorgan wastewater treatment plant in northern Iran). Cleaner Engineering and Technology, 4, 100210. https://doi.org/10.1016/j.clet.2021.100210
94. Mishra, V. K., & Tripathi, B. D. (2008). Concurrent removal and accumulation of heavy metals by the three aquatic macrophytes. Bioresource Technology, 99(15), 7091–7097. https://doi.org/10.1016/j.biortech.2008.01.002
95. Montes-Bayón, M., Yanes, E. G., Ponce de León, C., Jayasimhulu, K., Stalcup, A., Shann, J., & Caruso, J. A. (2002). Initial Studies of Selenium Speciation in Brassica juncea by LC with ICPMS and ES-MS Detection: An Approach for Phytoremediation Studies. Analytical Chemistry, 74(1), 107–113. https://doi.org/10.1021/ac0106804
96. Morant, M., Bak, S., Møller, B. L., & Werck-Reichhart, D. (2003). Plant cytochromes P450: Tools for pharmacology, plant protection and phytoremediation. Current Opinion in Biotechnology, 14(2), 151–162. https://doi.org/10.1016/S0958-1669(03)00024-7
97. Morikawa, H., Takahashi, M., & Kawamura, Y. (2003). Metabolism and Genetics of Atmospheric Nitrogen Dioxide Control Using Pollutant-Philic Plants. In Phytoremediation (pp. 763–786). John Wiley & Sons, Ltd. https://doi.org/10.1002/047127304X.ch25
98. Mustafa, H. M., & Hayder, G. (2021). Recent studies on applications of aquatic weed plants in phytoremediation of wastewater: A review article. Ain Shams Engineering Journal, 12(1), 355–365. https://doi.org/10.1016/j.asej.2020.05.009
99. Nedjimi, B. (2021). Phytoremediation: A sustainable environmental technology for heavy metals decontamination. SN Applied Sciences, 3(3), 286. https://doi.org/10.1007/s42452-021-04301-4
100. Newman, L. A., & Reynolds, C. M. (2005). Bacteria and phytoremediation: New uses for endophytic bacteria in plants. Trends in Biotechnology, 23(1), 6–8. https://doi.org/10.1016/j.tibtech.2004.11.010
101. O'Connor, D., Zheng, X., Hou, D., Shen, Z., Li, G., Miao, G., … Guo, M. (2019). Phytoremediation: Climate change resilience and sustainability assessment at a coastal brownfield redevelopment. Environment International, 130, 104945. https://doi.org/10.1016/j.envint.2019.104945

102. Oloumi, H., Mousavi, E. A., & Nejad, R. M. (2018). Multi-Wall Carbon Nanotubes Effects on Plant Seedlings Growth and Cadmium/Lead Uptake In Vitro. Russian Journal of Plant Physiology, 65(2), 260–268. https://doi.org/10.1134/S102144371802019X

103. Ouvrard, S., Leglize, P., & Morel, J. L. (2014). PAH Phytoremediation: Rhizodegradation or Rhizoattenuation? International Journal of Phytoremediation, 16(1), 46–61. https://doi.org/10.1080/15226514.2012.759527

104. Ouyang, Y. (2002). Phytoremediation: Modeling plant uptake and contaminant transport in the soil–plant–atmosphere continuum. Journal of Hydrology, 266(1), 66–82. https://doi.org/10.1016/S0022-1694(02)00116-6

105. Panz, K., & Miksch, K. (2012). Phytoremediation of explosives (TNT, RDX, HMX) by wild-type and transgenic plants. Journal of Environmental Management, 113, 85–92. https://doi.org/10.1016/j.jenvman.2012.08.016

106. Pavlostathis, S. G., Comstock, K. K., Jacobson, M. E., & Saunders, F. M. (1998). Transformaton of 2,4,6-trinitrotoluene by the aquatic plant Myriophyllum spicatum. Environmental Toxicology and Chemistry, 17(11), 2266–2273. https://doi.org/10.1002/etc.5620171117

107. Pinto, A. P., de Varennes, A., Dias, C. M. B., & Lopes, M. E. (2018). Microbial-Assisted Phytoremediation: A Convenient Use of Plant and Microbes to Clean Up Soils. In Abid A. Ansari, S. S. Gill, R. Gill, G. R. Lanza, & L. Newman (Eds.), Phytoremediation: Management of Environmental Contaminants, Volume 6 (pp. 21–87). Cham: Springer International Publishing. https://doi.org/10.1007/978-3-319-99651-6_2

108. Pittarello, M., Busato, J. G., Carletti, P., & Dobbss, L. B. (2017). Possible developments for ex situ phytoremediation of contaminated sediments, in tropical and subtropical regions – Review. Chemosphere, 182, 707–719. https://doi.org/10.1016/j.chemosphere.2017.04.093

109. Pradhan, S. P., Conrad, J. R., Paterek, J. R., & Srivastava, V. J. (1998). Potential of Phytoremediation for Treatment of PAHs in Soil at MGP Sites. Journal of Soil Contamination, 7(4), 467–480. https://doi.org/10.1080/10588339891334401

110. Prasad, M. N. V., & Tewari, J. C. (2016). Chapter 3 - Prosopis juliflora (Sw) DC: Potential for Bioremediation and Bioeconomy. In M. N. V.

Prasad (Ed.), Bioremediation and Bioeconomy (pp. 49–76). Elsevier. https://doi.org/10.1016/B978-0-12-802830-8.00003-4

111. Prasad, Majeti Narasimha Vara. (2019). Chapter 19—Prospects for Manipulation of Molecular Mechanisms and Transgenic Approaches in Aquatic Macrophytes for Remediation of Toxic Metals and Metalloids in Wastewaters. In Majeti Narasimha Vara Prasad (Ed.), Transgenic Plant Technology for Remediation of Toxic Metals and Metalloids (pp. 395–428). Academic Press. https://doi.org/10.1016/B978-0-12-814389-6.00019-5
112. Pulford, I. D., & Watson, C. (2003). Phytoremediation of heavy metal-contaminated land by trees—A review. Environment International, 29(4), 529–540. https://doi.org/10.1016/S0160-4120(02)00152-6
113. Radwan, S. (2009). Phytoremediation for Oily Desert Soils. In A. Singh, R. C. Kuhad, & O. P. Ward (Eds.), Advances in Applied Bioremediation (pp. 279–298). Berlin, Heidelberg: Springer. https://doi.org/10.1007/978-3-540-89621-0_15
114. Rai, P. K. (2008). Heavy Metal Pollution in Aquatic Ecosystems and its Phytoremediation using Wetland Plants: An ecosustainable approach. International Journal of Phytoremediation, 10(2), 133–160. https://doi.org/10.1080/15226510801913918
115. Rai, P. K. (2009). Heavy Metal Phytoremediation from Aquatic Ecosystems with Special Reference to Macrophytes. Critical Reviews in Environmental Science and Technology, 39(9), 697–753. https://doi.org/10.1080/10643380801910058
116. Reeves, R.D. (2006). Hyperaccumulation of trace elements by plants. In J.-L. Morel, G. Echevarria, & N. Goncharova (Eds.), Phytoremediation of Metal-Contaminated Soils (pp. 25–52). Dordrecht: Springer Netherlands. https://doi.org/10.1007/1-4020-4688-X_2
117. Reeves, Roger D., Schwartz, C., Morel, J. L., & Edmondson, J. (2001). Distribution and Metal-Accumulating Behavior of Thlaspi caerulescens and Associated Metallophytes in France. International Journal of Phytoremediation, 3(2), 145–172. https://doi.org/10.1080/15226510108500054
118. Reichenauer, T. G., & Germida, J. J. (2008). Phytoremediation of Organic Contaminants in Soil and Groundwater. ChemSusChem, 1(8–9), 708–717. https://doi.org/10.1002/cssc.200800125
119. Rentz, J. A., Alvarez, P. J. J., & Schnoor, J. L. (2005). Benzo[a]pyrene co-metabolism in the presence of plant root extracts and exudates:

Implications for phytoremediation. Environmental Pollution, 136(3), 477–484. https://doi.org/10.1016/j.envpol.2004.12.034

120. Rezania, S., Park, J., Rupani, P. F., Darajeh, N., Xu, X., & Shahrokhishahraki, R. (2019). Phytoremediation potential and control of Phragmites australis as a green phytomass: An overview. Environmental Science and Pollution Research, 26(8), 7428–7441. https://doi.org/10.1007/s11356-019-04300-4
121. Rizwan, M., Ali, S., Zia ur Rehman, M., Rinklebe, J., Tsang, D. C. W., Bashir, A., … Ok, Y. S. (2018). Cadmium phytoremediation potential of Brassica crop species: A review. Science of The Total Environment, 631–632, 1175–1191. https://doi.org/10.1016/j.scitotenv.2018.03.104
122. Rono, J. K., Le Wang, L., Wu, X. C., Cao, H. W., Zhao, Y. N., Khan, I. U., & Yang, Z. M. (2021). Identification of a new function of metallothionein-like gene OsMT1e for cadmium detoxification and potential phytoremediation. Chemosphere, 265, 129136. https://doi.org/10.1016/j.chemosphere.2020.129136
123. Ruiz, O. N., Alvarez, D., Torres, C., Roman, L., & Daniell, H. (2011). Metallothionein expression in chloroplasts enhances mercury accumulation and phytoremediation capability. Plant Biotechnology Journal, 9(5), 609–617. https://doi.org/10.1111/j.1467-7652.2011.00616.x
124. Rylott, E. L., & Bruce, N. C. (2009). Plants disarm soil: Engineering plants for the phytoremediation of explosives. Trends in Biotechnology, 27(2), 73–81. https://doi.org/10.1016/j.tibtech.2008.11.001
125. Sahu, O., & Singh, N. (2019). 13—Significance of bioadsorption process on textile industry wastewater. In Shahid-ul-Islam & B. S. Butola (Eds.), The Impact and Prospects of Green Chemistry for Textile Technology (pp. 367–416). Woodhead Publishing. https://doi.org/10.1016/B978-0-08-102491-1.00013-7
126. Sai Saraswathi, V., Kamarudheen, N., BhaskaraRao, K. V., & Santhakumar, K. (2017). Phytoremediation of dyes using Lagerstroemia speciosa mediated silver nanoparticles and its biofilm activity against clinical strains Pseudomonas aeruginosa. Journal of Photochemistry and Photobiology B: Biology, 168, 107–116. https://doi.org/10.1016/j.jphotobiol.2017.02.004
127. Salt, D. E., Smith, R. D., & Raskin, I. (1998). Phytoremediation. Annual Review of Plant Physiology and Plant Molecular Biology, 49(1), 643–668. https://doi.org/10.1146/annurev.arplant.49.1.643

128. Sarma, H., Nava, A. R., & Prasad, M. N. V. (2019). Mechanistic understanding and future prospect of microbe-enhanced phytoremediation of polycyclic aromatic hydrocarbons in soil. Environmental Technology & Innovation, 13, 318–330. https://doi.org/10.1016/j.eti.2018.12.004
129. Schröder, P., Daubner, D., Maier, H., Neustifter, J., & Debus, R. (2008). Phytoremediation of organic xenobiotics – Glutathione dependent detoxification in Phragmites plants from European treatment sites. Bioresource Technology, 99(15), 7183–7191. https://doi.org/10.1016/j.biortech.2007.12.081
130. Schwitzguébel, J.-P., Comino, E., Plata, N., & Khalvati, M. (2011). Is phytoremediation a sustainable and reliable approach to clean-up contaminated water and soil in Alpine areas? Environmental Science and Pollution Research, 18(6), 842–856. https://doi.org/10.1007/s11356-011-0498-0
131. Sharma, P. (2021). Efficiency of bacteria and bacterial assisted phytoremediation of heavy metals: An update. Bioresource Technology, 328, 124835. https://doi.org/10.1016/j.biortech.2021.124835
132. Sharma, S., Singh, B., & Manchanda, V. K. (2015). Phytoremediation: Role of terrestrial plants and aquatic macrophytes in the remediation of radionuclides and heavy metal contaminated soil and water. Environmental Science and Pollution Research, 22(2), 946–962. https://doi.org/10.1007/s11356-014-3635-8
133. Shen, X., Dai, M., Yang, J., Sun, L., Tan, X., Peng, C., … Naz, I. (2022). A critical review on the phytoremediation of heavy metals from environment: Performance and challenges. Chemosphere, 291, 132979. https://doi.org/10.1016/j.chemosphere.2021.132979
134. Sleegers, F., & Hisle, M. (2018). Redesigning Abandoned Gas Stations Through Phytotechnologies. In Abid A. Ansari, S. S. Gill, R. Gill, G. R. Lanza, & L. Newman (Eds.), Phytoremediation: Management of Environmental Contaminants, Volume 6 (pp. 3–20). Cham: Springer International Publishing. https://doi.org/10.1007/978-3-319-99651-6_1
135. Subhashini, V., & Swamy, A. V. V. S. (2017). Potential of Catharanthus roseus (L.) in Phytoremediation of Heavy Metals. In M. Naeem, T. Aftab, & M. M. A. Khan (Eds.), Catharanthus roseus: Current Research and Future Prospects (pp. 349–364). Cham: Springer International Publishing. https://doi.org/10.1007/978-3-319-51620-2_15

136. Suresh, B., & Ravishankar, G. A. (2004). Phytoremediation—A Novel and Promising Approach for Environmental Clean-up. Critical Reviews in Biotechnology, 24(2–3), 97–124. https://doi.org/10.1080/07388550490493627

137. T. Macek, K. Francova, L. Kochánková, P. Lovecká, E. Ryslava, J. Rezek, … M. Mackova. (2004). Phytoremediation—Biological Cleaning of a Polluted Environment. Reviews on Environmental Health, 19(1), 63–82. https://doi.org/10.1515/REVEH.2004.19.1.63

138. Tang, W., Wu, M., Lou, W., & Yang, C. (2021). Role of extracellular polymeric substances and enhanced performance for biological removal of carbonaceous organic matters and ammonia from wastewater with high salinity and low nutrient concentrations. Bioresource Technology, 326, 124764. https://doi.org/10.1016/j.biortech.2021.124764

139. Thomas, G., Sheridan, C., & Holm, P. E. (2022). A critical review of phytoremediation for acid mine drainage-impacted environments. Science of The Total Environment, 811, 152230. https://doi.org/10.1016/j.scitotenv.2021.152230

140. Tlustoš, P., Pavlíková, D., Száková, J., & Balík, J. (2006). Plant accumulation capacity for potentially toxic elements. In J.-L. Morel, G. Echevarria, & N. Goncharova (Eds.), Phytoremediation of Metal-Contaminated Soils (pp. 53–84). Dordrecht: Springer Netherlands. https://doi.org/10.1007/1-4020-4688-X_3

141. Van Ginneken, L., Meers, E., Guisson, R., Ruttens, A., Elst, K., Tack, F. M. G., … Dejonghe, W. (2007). Phytoremediation for heavy metal-contaminated soils combined with bioenergy production. Journal of Environmental Engineering and Landscape Management, 15(4), 227–236. https://doi.org/10.1080/16486897.2007.9636935

142. Vangronsveld, J., Herzig, R., Weyens, N., Boulet, J., Adriaensen, K., Ruttens, A., … Mench, M. (2009). Phytoremediation of contaminated soils and groundwater: Lessons from the field. Environmental Science and Pollution Research, 16(7), 765–794. https://doi.org/10.1007/s11356-009-0213-6

143. Venkatachalam, P., Jayaraj, M., Manikandan, R., Geetha, N., Rene, E. R., Sharma, N. C., & Sahi, S. V. (2017). Zinc oxide nanoparticles (ZnONPs) alleviate heavy metal-induced toxicity in Leucaena leucocephala seedlings: A physiochemical analysis. Plant Physiology and Biochemistry, 110, 59–69. https://doi.org/10.1016/j.plaphy.2016.08.022

144. Victor, K. K., Séka, Y., Norbert, K. K., Sanogo, T. A., & Celestin, A. B. (2016). Phytoremediation of wastewater toxicity using water hyacinth (Eichhornia crassipes) and water lettuce (Pistia stratiotes). International Journal of Phytoremediation, 18(10), 949–955. https://doi.org/10.1080/15226514.2016.1183567

145. Wang, J., Zhang, Z., Su, Y., He, W., He, F., & Song, H. (2008). Phytoremediation of petroleum polluted soil. Petroleum Science, 5(2), 167–171. https://doi.org/10.1007/s12182-008-0026-0

146. Wang, Liang, Lin, H., Dong, Y., Li, B., & He, Y. (2020). Effects of endophytes inoculation on rhizosphere and endosphere microecology of Indian mustard (Brassica juncea) grown in vanadium-contaminated soil and its enhancement on phytoremediation. Chemosphere, 240, 124891. https://doi.org/10.1016/j.chemosphere.2019.124891

147. Wang, Liuwei, Hou, D., Shen, Z., Zhu, J., Jia, X., Ok, Y. S., … Rinklebe, J. (2020). Field trials of phytomining and phytoremediation: A critical review of influencing factors and effects of additives. Critical Reviews in Environmental Science and Technology, 50(24), 2724–2774. https://doi.org/10.1080/10643389.2019.1705724

148. Ward, O. P., & Singh, A. (2004). Soil Bioremediation and Phytoremediation—An Overview. In Ajay Singh & O. P. Ward (Eds.), Applied Bioremediation and Phytoremediation (pp. 1–12). Berlin, Heidelberg: Springer. https://doi.org/10.1007/978-3-662-05794-0_1

149. Watanabe, M. E. (2011, June 8). Phytoremediation on the Brink of Commericialization. https://doi.org/10.1021/es972219s

150. Williams, J. B. (2002). Phytoremediation in Wetland Ecosystems: Progress, Problems, and Potential. Critical Reviews in Plant Sciences, 21(6), 607–635. https://doi.org/10.1080/0735-260291044386

151. World Health Organization. Regional Office for Europe. (2021). Mercury and human health: Educational course. World Health Organization. Regional Office for Europe. Retrieved from https://apps.who.int/iris/handle/10665/345443

152. Yan, Q., Gao, X., Guo, J., Zhu, Z., & Feng, G. (2016). Insights into the molecular mechanism of the responses for Cyperus alternifolius to PhACs stress in constructed wetlands. Chemosphere, 164, 278–289. https://doi.org/10.1016/j.chemosphere.2016.08.103

153. Yang, X., Feng, Y., He, Z., & Stoffella, P. J. (2005). Molecular mechanisms of heavy metal hyperaccumulation and phytoremediation.

Journal of Trace Elements in Medicine and Biology, 18(4), 339–353. https://doi.org/10.1016/j.jtemb.2005.02.007

154. Zitka, O., Krystofova, O., Hynek, D., Sobrova, P., Kaiser, J., Sochor, J., … Adam, V. (2013). Metal Transporters in Plants. In Dharmendra K. Gupta, F. J. Corpas, & J. M. Palma (Eds.), Heavy Metal Stress in Plants (pp. 19–41). Berlin, Heidelberg: Springer. https://doi.org/10.1007/978-3-642-38469-1_2

INDEX

B

C

D

E

F

G

H

I

K

L

M

Q

R